LETTERING ON CERAMICS

Mary White

A&C Black • London

American Ceramic Society • Ohio

First published in Great Britain 2003
A&C Black Publishers
37 Soho Square
London W1D 3QZ
www.acblack.com

ISBN 07136-6264-6

Published simultaneously in the USA by
The American Ceramic Society, 735 Ceramic Place,
Westerville, Ohio 43081, USA
www.ceramics.org

ISBN 1-57498-216-8

CIP records for this book are available from the British Library and the US Library of Congress.

Cover illustration: Bowl by Mary White, diameter: 28 cm (11 in.), from a private collection (U.S.A.).

Frontispiece: Porcelain teapot with cut-out letters by Mary White. Inside is a crater with coloured glass at the bottom. A 'fun' piece – it cannot pour. Height: 27 cm (10⅝ in.). By Mary White.

Page 6: Piece by Anneke Harting.

Cover design by Dorothy Moir.

Printed and bound in Singapore by Tien Wah Press.

A & C Black uses paper produced with elemental chlorine-free pulp, harvested from managed sustainable forests.

Contents

禪

Chapter One
Brief History

It is said that history began with the ability to communicate. Primitive man was not stupid, he was inexperienced. He must have possessed an inventive and resourceful brain to take the first steps leading to our vast knowledge today. We know he was artistic; we have proof of this in the cave paintings that have a sense of design and a skill that modern artists today envy. When he discovered that he could make sounds that other beings could understand he developed a language of communication. Eventually, as his knowledge and abilities increased he found it necessary to record his language in a visible way.

The story of writing began between the Euphrates and Tigris rivers in Mesopotamia. From the 6th to the 2nd millenium BC this area of the Middle East (between the Persian Gulf and Baghdad) consisted of two regions: the land Sumer in the south and the land Akkad in the north. Their languages were quite different. The most civilised members of society congregated in towns such as Babylon, and were ruled and protected by aristocrats, surrounded by court officers, priests, traders, farmers and labourers. Historians tell us that writing began with accountancy – the necessity to record sales of slaves, property, areas of land, etc. The earliest form of writing on clay was found near the ancient temples and seemed to be lists of cereals and animals – the earliest book-keeping! Some showed lists of farmers, slaves, bakers, smiths. They show that people already had some form of money and also lent it and demanded interest.

The first examples of this writing were stylised symbols: these are called 'pictograms'. Man's artistic skill enabled him to invent small pictures to represent single words: examples of these early records have been found in the Near East, and the Sumerians developed a pictographic script in the late 4th millenium BC. The Egyptians used the same ideas and formed their own concept of 'hieroglyphs'. Early writing was carved in stone or painted on walls, but a quicker, more personal method was necessary. People were familiar with clay and they had discovered that it could be baked in fire, so it was a natural development for them to make marks on damp clay.

There were about 1,550 different signs; these became more complicated and are now called 'ideograms'. In the beginning signs were made with reeds cut to a narrow triangle one end and a small round shape the other end. The tools became more sophisticated and were cut in metal or bone, making elongated triangular marks. The clay tablets were made to be carried in a pocket or held in the hand. It is thought that apprentice scribes prepared the tablets, which could be small or as large as 30 cm (11¾ in.) square. Usually one side was flat and the other curved so that the flat side was not damaged by pressure when the other side was written upon. The tablet could be dried out or later moistened, smoothed and reused, or it could be fired and made permanent. Schools existed to train young scribes, and school tablets have survived with the instructor's work on one side and the student's on the other. A trained scribe could be employed as a secretary in prestigious official or religious work, or in trade or agriculture. Writing was started at the top left-hand corner of the tablet in a column moving downwards, then the next column to the right moving similarly. The scribe would turn the tablet over from the bottom and work in reverse columns, right to left. The finished tablets were intriguing pieces of design and are called 'cuneiform'.

'Cuneiform', clay tablet showing early writing in Sumerian, *c.* 3000–2600BC. *Photograph © The British Museum.*

'Cuneiform', clay tablet with early writing, this time in the form of raised hieroglyphs. From the site of Tellumar near Baghdad, *c.* 14–12th cenury BC. *Photograph © The British Museum.*

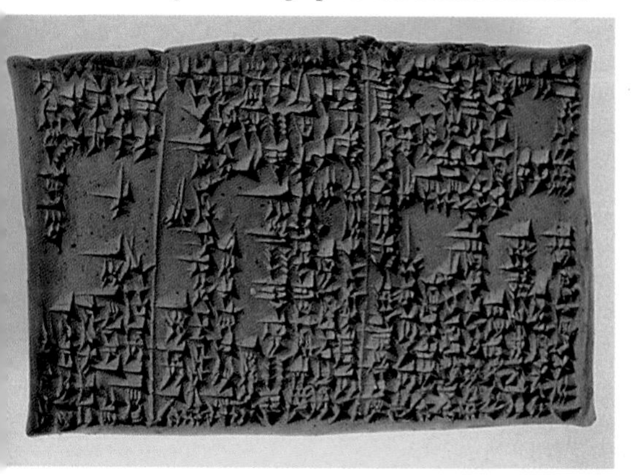

Seals were often made in clay and used as marks of ownership. Clay has the great advantage that it can be made fireproof and waterproof, so that archaeologists have found thousands of tablets that survived conquests and conflagrations.

Eventually the number of signs was reduced and gradually the alphabet was introduced; it developed from many sources in the Middle East, but the Phoenicians are thought to be the first to originate it, around 1000 BC. It came into use through the Greeks but we do not know when and how. The Phoenicians were great traders and travelled the world. They left little information, but they used an alphabet of 22 consonants. Most Phoenician inscriptions originate in Phoenicia, Cyprus and Carthage. After the destruction of Carthage in 146 BC, the script was used in North Africa and disappeared in the 2nd century AD. Eventually clay tablets fell out of use because the Egyptians developed papyrus as a writing material and in the 2nd century BC, paper was invented in China. Throughout the centuries, scribes were able to use other materials and tools for writing; however, letters continued to be used on pottery as a decoration. In the 8th century BC, impressive lettered Greek vases were found in Athens, and there are many Hebrew inscriptions which have survived on pottery, showing the

Dish by Thomas Toft, 1671. The letters and cross-hatching around the edge have been applied with a slip-trailer. *Photograph courtesy of Grosvenor Museum, Chester.*

most developed cursive writing of the 8th century BC. In the same century Greek vases were made with lettering scratched through a coloured slip to the natural clay below, but later in the 6th century BC, letters were painted. In the 13th century AD many very beautiful Persian vases were made with fine inscriptions completely integrated into the design. In the 18th century, rather crude letters were used on majolica in Italy, and in Britain. Thomas Toft became known for his lettered slipware. He made large plates, often 43–56 cm (17–22 in.) in diameter and 7 cm (2.8 in.) deep, trailing elaborate line designs and simple lettering. He used red clay covered with white slip and drew his design with a dark brown slip in a pipette, like icing from an icing-bag onto cake. He varied the colour with a lighter, redder tone. Dots of white slip gave life and sparkle. A typical design motif was lattice on the rims.

Today ceramic artists are beginning to realise the potentials of using the alphabet in their work. A revolution is taking place – letters are no longer purely functional, they have become an art form in themselves and not necessarily readable. Especially in the last decade, calligraphy has been used in painting and painting in calligraphy. The increased use of calligraphy in ceramics has nothing to do with 'souvenirs' or 'presentations' – it is an expression of feelings and imagination, an exciting art form. A new era of 'letters on clay' is beginning.

Calligraphic strokes can express feeling and ideas in the same way as painting or writing, and not only on paper. For example, when I started my series of 'Calligraphic Marks' bowls, I tried to capture the excitement of the way the bowl was created on a fast-rotating wheel, the strokes flowing with the movement. The marks are part of the object and were created by making it.

Chapter Two
Tools and Materials

Tools

These are some of the tools I find useful.

Band wheel, this is a small revolving stand which enables you to work on something from all sides without having to move the piece itself. It is very useful for lettering all the way round an object.
Brushes, ranging from old ones for mixing up colour and slip roughly, to finely pointed sables (starting from size 00) to do fine brushwork on pieces.
Calico or some other material is good for rolling out clay on.
Knives and **scalpels**, for cutting card, paper and clay. These should be sharp in order to get a clean edge.
Nibs for lettering, having a selection of widths means you can choose the correct size to suit the work. These don't have to be specifically lettering nibs, many other tools will work well. For instance screwdrivers have a flat tip, are very strong and can be used to 'write' on the surface, removing the surface of the clay and leaving the letter engraved. You can also try wood engraving or lino-cutting tools, pieces of balsa wood, or make your own specifically.
Paper and **card** are useful for cutting out letters to use as guides, or to use as resists while decorating.
Plaster batt to work on, can be bought from a pottery suppliers. This helps to dry clay out if it is very wet, but be careful it doesn't dry the work out too much or too fast.
Plastic buckets, with lids, preferably! These are useful for keeping scraps of clay in, or glazes.
Plastic sheeting to cover your work and stop it drying out. You could use plastic from the dry cleaners or even soft plastic bags are perfectly adequate.
Rolling pin for rolling slabs, this really must be wooden or the clay will stick to it. One without handles, that is, a simple smooth rolling pin will be easier to use.
Ruler, **set square** and **pencils**, for designing your letters and clay forms.
Smooth board is also useful to work on, and also to keep finished pieces on while they dry. Avoid plywood as the dampness in the clay will warp it.
Wooden slats to be used as guides when rolling out slabs of clay. These must be the same thickness otherwise the slab will be uneven. These can be obtained from a pottery suppliers, but any timber yard and some DIY shops will be able to cut you two strips of smooth wood. To start off, try 45 cm (17 ¾ in.) long and about 8 mm (5/16 in.) in width. As you become more proficient, you may want different thicknesses.

Clay

Clay is not a 'soil', although the two are related. The soil we cultivate in our garden is a mixture of clay, sand and decayed vegetable matter which we call 'humus'. This is not usually a very thick

Newel Post by Mary White. Stoneware, slab-built. Lettering made with calligraphy nib and partly gilded wih lustre. Height: 51 cm (20 in.). In a private collection. *Photograph by Mary White.*

layer, but in some areas there is a subsoil which can vary in thickness – this is clay. The great advantage of clay is that it is often plastic and will hold its shape so that it can be formed into vessels or sculptures, and fired. Garden soil will not. The study of clay composition is complicated as it derives from particles of rock, but some clays are more plastic than others, and they fire at different temperatures. Today we are fortunate to have ceramic suppliers who stock ready prepared and tested clays – we only need to study their catalogues and buy what we need. Basically we can divide pottery clays into stoneware, fireclay, and earthenware. Porcelain is a manmade 'body' composed of various natural ingredients.

Stoneware

Usually plastic, with a wide firing range from cone 6 to cone 10. The colour changes when fired and can vary from off-white, to buff, to grey, and black.

Fireclay

A high-firing clay used mostly for insulating and building kilns and making kiln furniture. Some are fine enough to throw, but usually they are too coarse.

Earthenware

These are low-firing clays maturing between cones 08 and 02. Usually they contain iron oxide so they are brown in colour and fire to warm terracotta, but they can be white. They are pleasant to work with as they are very plastic, but because of the low firing, pots remain porous to liquids unless glazed inside with a hard glossy glaze. These glazes often contain lead, so are dangerous to health if in contact with acidic drinks such as wine or orange juice (safe lead glazes have now been developed which can be bought from suppliers or made by potters). Potters can also make their own glazes which do not contain lead compounds. A simple red or white earthenware clay is a good clay to start with. Ask your pottery supplier for advice if unsure.

Porcelain

These bodies are made from kaolin, feldspar, and flint, sometimes with added ball clay. Porcelain is mainly used for commercial wares produced in plaster moulds or by 'jiggering', but for the last century it has become a more popular material for ceramic artists as it has a unique delicacy, fires white and can be translucent – these qualities are inspiring. With an increasing knowledge of glazes, wonderful results can be achieved and fine sculptural forms can be fired. However porcelain shrinks more than stoneware when fired, so it is not easy to use the two clays in combination.

Slip

This is a mixture of clay and water, sometimes with oxides added for colour. Slips can give a smooth texture to a coarse body, or can change the colour of the body, giving the opportunity of enriching the surface design. It is not easy to find the right clay for slipware as it must be plastic and smooth and fire at the same temperature as the main body. Often the safest option is to use the same clay as is used for the main body, with stains or oxides added to alter the colour. The slip clay is dried, broken or cut into pieces, ground to a powder, preferably in a mortar. This is then transferred into a bucket, hot water added and left once a week, with a daily stirring. Eventually this mixture must be pushed through a 120s mesh sieve, with the final slip taking on the consistency of a slightly thicker pancake mixture. According to the clay you use you can make red, brown, black or blue slip.

Glazes

Some of the ceramic artists featured in the 'Gallery' section of the book have contributed special glaze recipes. I find this very generous as often glazes are treasured secrets. There are, however, many good books with recipes and illustrations available today, and all you need to do is experiment! Often sculptural work does not need a glaze, and it is also possible (for some types of work) to polish it with a good wax polish!

Colours

If you are using two clays which will fire different colours (e.g. red and white earthenware), but will fire to the same temperature then you can make a slip of one of them and use it for your colour. You can mix powdered oxides with water and use these, but this is not easy and will require some practice and experimentation.

Fortunately today there are ready-mixed colours for painting either under a glaze or over glaze and they are so well mixed that they will give you a good idea of how they will look when fired. These are available in solid form like colours in a box of watercolours, all you do is moisten them with a brush and water and then paint with them. You can also buy them in a liquid form in small jars, but you must be careful to replace the lids or they will dry out! These can be bought either to use directly on dry or biscuit-fired clay (as under-glaze colours) or to paint on pieces that have already been glazed (onglaze colours).

Glazing can be tricky, but if you buy a ready-made glaze to start with you can get some good results fairly easily. There is not enough space here to give a full account of glazing, but there are many good books on the subject!

Chapter Three
Methods

Brush

Basically there are three ways to design letters on clay: on the surface, into the surface and built-up on the surface. Most of my work has been done with a brush, the size varying with the design. I find size 0 or 00 suits me best, but sometimes I use flat brushes of various sizes which can produce letters much like working on paper. Always use the best sable, cheaper brushes have not got the same spring, but as the clay and the colours are abrasive, brushes do not last long, so they should be carefully cleaned after use and soaped to retain their shape. To 'soap' a brush, carefully draw the brush over a moist tablet of soap *in one direction only*, and then gently smooth it into shape and leave to dry. When you want to use it the soap will come out with a gentle washing. I keep my old brushes to use for mixing and if I need a palette I find the smallest possible to avoid waste. Plastic egg containers from an old fridge make very useful palettes. Potters' suppliers have underglaze and overglaze ready mixed either in small jars in liquid form or in small blocks in cases just like watercolours. We are fortunate these days that we do not have to mix our own colours using basic oxides with endless experiments to make the right colour – today

RIGHT, TOP TO BOTTOM Lidded box; abstract form; rounded box; Mary White. The letters on these forms have all been hand-painted using brushes and oxides.

Bowl by Mary White. Showing brushstrokes. Width: 25 cm (9⅞ in.). *Photograph by Inge Mess.*

the ready-mixed colourants are even coloured to show what they will be like when fired!

Underglaze can be used on dry or leatherhard clay before biscuit firing, or when you have biscuit fired the work. When glazing I prefer to spray as then there is not so much risk of smudging your letters. Onglaze colours must be fired at the temperature stated in the catalogue. Calligraphers will be pleased to learn that it is possible to use gold on some letters. This is bought in a very small, and expensive, bottle which should be turned over several times before opening to use and must also be fired at the recommended temperature. The fine gold is suspended in a dark gummy liquid and you must be extremely careful not to get any spots on your work. Be specially careful that your hand or fingers do not touch the finished work as it remains sticky for some time and if you mark the pot you cannot get it out! I have tried with turps and several spirits but a purple stain always remains after firing. It is possible to cover the finished work, when quite dry, with a fine handkerchief, but even that is risky – the gold could stick to it! Lettering a globe in gold is always nerve-wracking!

I repeat – you must keep experimenting – but always record what you do in a special notebook, not on scraps of paper because you will lose them – as I know from experience! Ideally always make small samples of what you are doing and write the details on the reverse with black oxide. I find the benefit of working on biscuit is that you can roughly pencil your design before you paint it, as pencil marks fire out. It is easy to correct your work (but not gold) on dry clay by softly scraping or using very fine sandpaper, but it is also possible on biscuit if you are careful. You can also use a fine brush dipped in water and then wiped almost

ABOVE Bowl, Mary White. The bowl has been decorated with a sponge cut into the shape of an A, as well as painted with gold lustre on top. Diameter: 8 cm (3⅛ in.).

BELOW Round bottle, Mary White. This rounded bottle form has been decorated using a sponge cut into the shape of an oval and dipped in slip. Dimensions: 15 x 12 cm (6 x 4¾ in.).

dry to clean-up or alter colour. Be very careful with this part as the colours are mostly poisonous, don't be tempted to suck the end of the paintbrush! These methods are only for correcting letters made with oxides.

Sponges

Fine sponges, especially washing-up sponges, are useful for colouring background areas. Paint the colour on the sponge with a large brush or dip it into colour mixed up in a small palette. You need to practice to achieve an all-over cover.

Sponges can also be cut with a very sharp knife into letters. Choose a bold, large style without serifs and remember to reverse them. It helps to design on paper or thin card first, and cut these out accurately, then reverse them, and place them on the sponge, using them as a guide to cut around. You could even use a biro to mark around the sponge and then cut on these lines. You need a thick piece of sponge (washing-up ones are ideal) to give support to the letters and use a very sharp, thin blade such as a scalpel or a craft knife and make the cuts about a quarter of an inch deep. Be careful not to undercut the letters, cut at right-angles or with a slight bevel outwards to give more strength to the letters. To print with these either paint the colour on the sponge with a brush or dip into colour mixed in a small palette as described above. One advantage of sponge is that it is pliable and can be used to stamp letters on a curved form. You can use it with underglaze colours on raw or biscuit clay and even with onglaze colours on the finished pot.

Stencils and resists

These can be made from waxed paper or even newspaper (which I prefer). The letters can be sponged through, stippled with a stiff brush, or sprayed. The difficulty with stencils is that the inner parts of the letters fall out when you are cutting them, but the stencils should be laid on leatherhard or damp clay so that they stick in place and this makes it easy to carefully replace the bits inside the letters. One of the advantages of using newspaper is that it easily wraps around a curved surface. Remember to make a photocopy of your design in case you need it. I think my photocopier is invaluable: I make copies of everything and file them.

I find an alphabet (or several alphabets) cut out singly in paper and in thin card is very useful. If treated with care they can be kept for repeated use and their virtue is that they can be arranged on the clay with freedom and then sprayed or brushed over with slip or glaze. A card alphabet can be kept as a guide for replacing broken letters and for cutting letters in newspapers (which are very useful for sticking in place with water on an upright or curved surface).

Cutting out a letter in paper using a scalpel.

Cut out letters from paper to be used as a resist for decorating.

Bowl, Mary White. Decorated using resist letters and the background filled in. The resist was then removed and brushstrokes added to the blank areas of letters. Width: 15 cm (6 in.). *Photograph by Inge Mess.*

Potters sometimes use wax as a resist, especially on the bottom of pots and around lids to prevent sticking and this method can also be used for painting letters with thin wax before glazing, but it is a method needing skill and constant use so I do not advise it for the inexperienced. However we are fortunate these days that there are many products on the market that we can use more easily. It is possible to buy special wax that is heated and can then be painted on while fluid. Calligraphers are familiar with modern resists and can experiment with these easily. Personally I often use a rubber solution (try Copydex) which can be rubbed off when dry if necessary. The big problem with resist as we all know, is that it ruins your brushes! You must clean them with the appropriate medium as soon as possible and try not to use your best sable ones.

Applying colour

Coloured slip or coloured oxides can be applied to the clay, perhaps with a large soft brush (mop) while the pot is revolving on a wheel or a bandwheel, or areas can be painted or sprayed. It is effective to scratch through the colour when it is drier, either with a sharp point or with an old lettering nib. With a nib you can actually write the same way that you would on paper, but instead of applying colour you scratch off the colour with your pen, revealing the clay underneath. With practice you can achieve the same freedom that you can with a pen on paper, with the great advantage that you can, with care, make corrections! I often use a screwdriver instead of a pen, and with imagination you can use other improvised tools such as strips of balsawood. You can make slip from the clay you are using with added oxides, or a clay of a different colour but the same firing temperature. For small quantities I simply powder some dry clay and mix water with it to make a pancake-like mixture which can be coloured with oxides or commercially ready-made underglaze colours. It is advisable to use a chemistry balance and make a record of the exact proportions of colour to clay. I usually make a number of tests, small rectangles of slabbed clay with a hole in the top to enable them to be hung up with string when fired. I paint or dip the slip on one side and write the information on the other with underglaze.

Slip-trailing

If you study peasant pottery in museums you will find many examples of slip-trailed work. Thomas Toft is the most well-known potter who worked in this way in the 17th century. Pottery suppliers usually stock a slip-trailer, but you can improvise if you find a plastic bottle with a fine nozzle, or even a dropper from a medicine bottle. Be prepared to practice, using different thicknesses of slip which you can make from different clays of the same firing temperature or mix colouring oxides with the clay you are using. Experiment to find out when you should use the slip so that it does not fall off as it dries: you have to work deliberately on damp clay. Use your slip-trailer the same way as icing the letters on a birthday cake. I have even used a slip-trailer filled with thick poster colour to work on paper, which was an invigorating experience as it was necessary to work extremely fast. I can recommend this as a very good way to practice before working on clay! Fill your trailer with colour then hold it at an angle to your work and press the slip out as you write the letters.

Tools, brushes, underglaze and onglaze colours. *Photograph by Inge Mess.*

Slab with signatures of calligraphers at Art in Action 2000 scratched in a black slip. *Photograph by Inge Mess.*

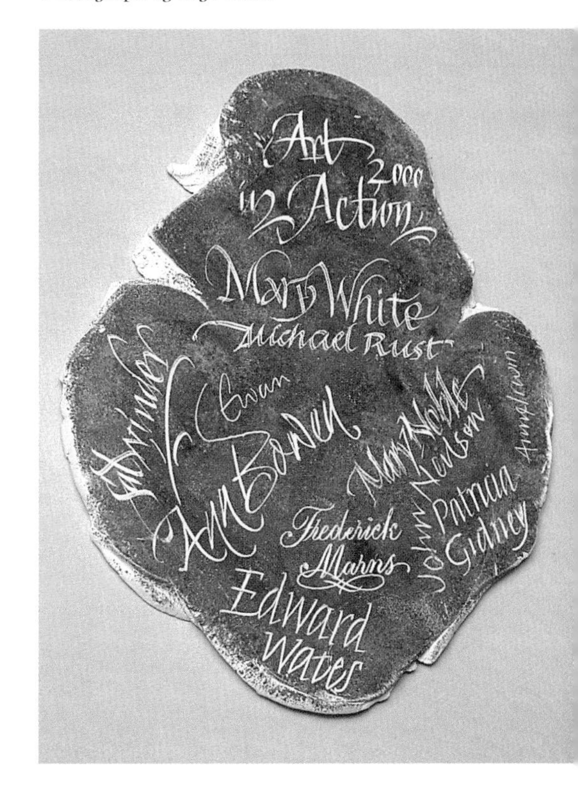

See the dish by Thomas Toft on p.9 and the pieces by Mary Wondrausch on pp. 79–80 for examples of slip-trailed letters.

Impressed and engraved letters

One of my favourite ways of using letters on clay is by impressing them into the surface. The principle is very much like printing with lead type only I use porcelain for the letters and they are pressed into the clay. This is very interesting when done on porcelain and left unglazed or treated with a thin transparent glaze as there is a fascinating play of light and shade. Alternatively the depressions can be filled in with oxide and the surface left clean. I usually use porcelain for these letters as it is such a fine clay that the letters made with it are sharp, but a good earthenware would also be suitable.

Rounded bottle form with pressed letters filled with oxides by Mary White.

First make the shanks by cutting out a shape that is convenient to hold when you are using the letters. Tap the top to make it quite flat. The letter must be carved in reverse on the top. To simplify this I draw the letter on a piece of paper or tracing paper using a biro. You might have to try different pens until you find one that will transfer your reversed drawing to the clay when rubbed gently onto the surface. The clay, of course, must be leatherhard. The next procedure is to carve the letter leaving it standing out like a piece of printing-type. Cut around the letter with a sharp, small knife, cutting with the blade held slightly bevelled outwards. This is very precise work, especially when you have to take out clay from inside the letters, but it is possible with patience and practice. Trim the top of the shank so that you only have a small area of clay around the letter. In a workshop we never have enough time to fire these letters, so we dry them in the heat or sun so that we can use them quickly. This does work if you are in a hurry, but they can get easily damaged, so it is advisable to bisque fire them, this will make them a bit smaller, so you must allow for shrinkage when you design. I rarely use these letters with colour painted on them: they must be kept clean.

Letters can also be carved out of rubber erasers and these could be used either to press into the clay or to be painted with colour and printed on the surface instead of into it. These can be washed clean and re-used.

Naturally one could use real metal type or the wonderful old wooden typefaces that you might be lucky enough to find in a flea-market.

Vase, by Mary White. Letters scratched into the porcelain with a lettering nib, giving shadow effect. Outside unglazed. Dimensions: 17 x 8 cm (6⅝ x. 3⅛ in.). *Photograph by Mary White.*

Porcelain letters to press into clay, carved out as stamps. *Photograph by Inge Mess.*

A trial piece engraved into the clay using a broad pen, or a screwdriver. *Photograph by Mary White.*

Raised letters

A favourite method for me is to keep cut-out alphabets in thin card to use as patterns for cut-out letters. These can be used in several interesting ways, varying from letters applied to ceramics to free-standing sculptures and on a very large scale to sculptures made entirely from cut-out letters. It is a region that is waiting to be explored.

For beginners it is recommended to start by including cut-out letters on your designs for tiles, then you can advance to adding them to built-up forms such as cubes and box forms, if you are an experienced potter or sculptor however, you can experiment with adding letters to any form.

To create raised letters, roll out clay of different thicknesses and using your card alphabet carefully cut around the letters keeping the angle of the knife consistent, either held at a right-angle or slightly sloping to form a bevel. I prefer a bevelled edge to the letters, but when the clay is rolled thinly it is better to keep to a right-angle. These letters must be kept in plastic until you are ready to fix them in place on your ceramic. You can use your card alphabet to try different arrangements on the clay and then either use those or the cut-out letters to mark the place you want them. To fix the letters in position you must score the undersides, and where they are to go on the clay, and then brush on a slip made from the same clay. Be careful placing them into

Detail of a panel with raised letters, by Mary White. *Photograph by Mary White.*

Letters cut out from card to be used as guides. *Photograph by Inge Mess.*

Letters cut out from porcelain to be impressed into clay or stuck onto work. *Photograph by Inge Mess.*

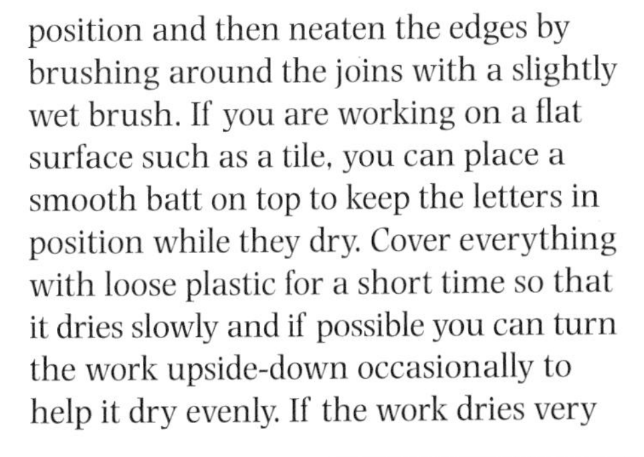

position and then neaten the edges by brushing around the joins with a slightly wet brush. If you are working on a flat surface such as a tile, you can place a smooth batt on top to keep the letters in position while they dry. Cover everything with loose plastic for a short time so that it dries slowly and if possible you can turn the work upside-down occasionally to help it dry evenly. If the work dries very unevenly or too quickly, the letters are liable to crack or fall off. If you are working with earthenware you could use two different colour clays of the same temperature.

A variety of tools used for cutting card, and cutting, smoothing and shaping clay. *Photograph by Inge Mess.*

Transfers or decals

It is possible to buy ready-made decals with all sorts of pictures, words and letters on, which can give the appearance of being printed onto the ceramic. These are placed in warm water, causing the backing paper and the thin plastic sheet to separate from one another. The plastic sheet with the letters or image on is then positioned onto the glazed surface of the object. The glaze allows the plastic sheet to be slid into position, and then it can be gently stuck down using a rubber kidney or finger to squeeze out the water from underneath, so that it lies flat against the glaze. The whole thing is then fired again so that the plastic sheet burns off, leaving the letters fused to the piece of work.

It is possible to print your own images onto decal paper using oil-based ceramic inks. Linocuts, lithography and screen printing are the simplest methods for this. The image is printed onto specially gummed decal paper, and used in the same way as the ready-made ones. It is even possible to print images from laser printers and photocopiers onto traditional decal paper using iron-rich toners. The paper and image (from fused iron oxide toner) is coated with a special varnish called Covercoat. Once dry, it becomes a thin plastic sheeting with the image on, creating the transfer. Placed in warm water again the plastic and image become softened and can be removed from the paper, and applied in the usual way. Using commercial decals is one of the simplest ways of printing letters onto ceramics, however printing onto ceramic is a vast area in itself with many other techniques. For further information see *Ceramics and Print* by Paul Scott (see bibliography).

Mixed techniques

When you have tried these different methods of using letters on clay, you can design with a combination of techniques. Ideas will come to you as you work, but always keep detailed records of what you do because it is easy to forget. Photographs of your work at different stages are also invaluable.

An example of various techniques built up on top of each other; resist letters, with sponged and painted letters on top.

A trial slab using various techniques.

Chapter Four
FORM

Slab

This is a method that is vey popular today with ceramic artists as it is versatile, does not need a wheel and can be built up into sculptural sizes. A variety of clays can be used, and as long as they shrink to the same size and fire to the same temperature, they can be mixed. Slabs can also be applied to wheel-thrown work to make imaginative forms. Ideally one needs a slab roller, which makes work much easier, as the clay comes out like pasta! But they are expensive and not every art school can afford one, although equipment is getting cheaper. To roll a slab yourself you need a wooden rolling-pin, a very level and smooth surface to work on and narrow strips of wood in pairs of different thicknesses from about 4 mm (⅛ in.) and 3 or 4 cm (1 ⅛–1 ½ in.) wide. I find a good general thickness is 8 mm (⅝ in.) and 45 cm (17 ¾ in.) long. These can be bought cheaply from a pottery suppliers or from a wood suppliers in a supermarket and should be made from a strong smooth wood that does not warp.

The working surface is ideally a plaster slab that you can make yourself or buy from a pottery suppliers. A plastic table top is good, but you must put a piece of smooth cloth over it or the clay will stick to the surface. Wood is possible if the grain is fine, but plywood will not last long as it warps when damp. Cover the wood with a piece of calico.

To make the slab, you need to place a block of wedged clay onto your working surface and thump it down with your fists or rolling-pin. Turn it over and repeat. Do this until the clay is nearing the thickness you need, then place the wooden guides one each side of the clay and begin to roll it resting the rolling-pin on the guides to control the eventual thickness, changing the direction at times. It is better to start in the middle, rolling outwards, then reverse the direction, again from the middle, then from top to bottom. The clay should eventually be quite level and ready for use.

If it is not enough for your proposed ceramic, carefully place it on a board

Sculptural coffee pot by Mary White, entirely constructed from slabs.

covered with plastic sheet and cover it with more plastic. This will stop it drying out while you make the next piece.
If you think you need several slabs you can stack them between boards in this way until wanted. With these slabs you can then make tiles, boxes, cubes, wall-panels or sculpture.

Slab lends itself very well to sculpture as you can design even complicated forms in paper or card in advance and join them in the usual way. The size depends only on the size of your kiln, but if you are very ambitious you can make pieces that fit together when fired.

OPPOSITE *Summer* by Mary White, slab-built sculptural form with painted letters. Height: 28 cm (11 in.). *Photograph by Mary White.*

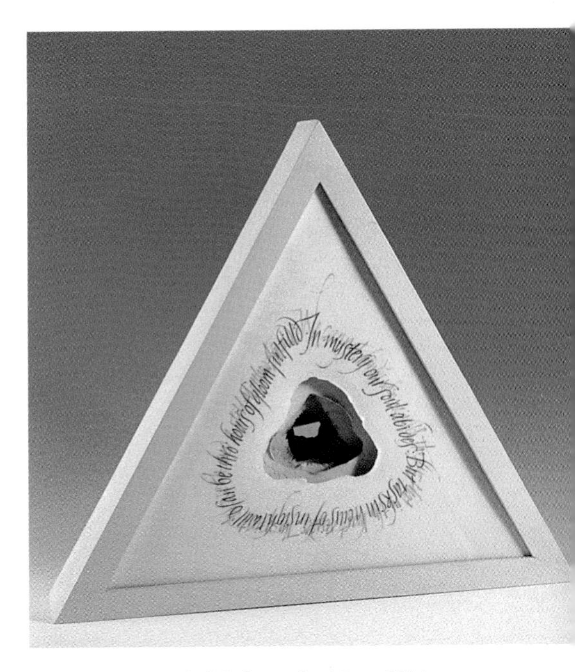

ABOVE Sculptural slab form, by Mary White. Dimensions: 21 x 21 cm (8¼ x 8¼ in.).

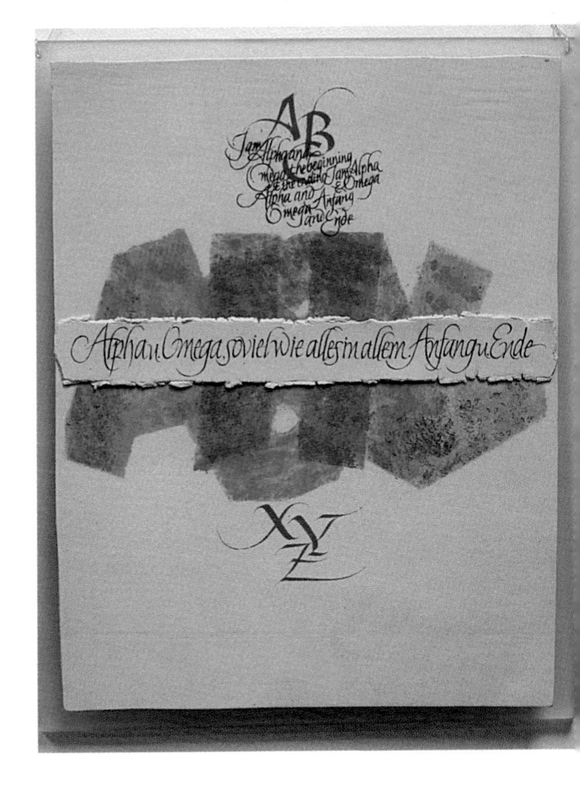

RIGHT *Alpha & Omega* by Mary White. I often use this wording in calligraphy as it lends itself well to design and can be written in any language. I used it on a small panel illustrated in the colour slide. First I cut large stencils for the separate letters of alpha and then used a coarse sponge to stencil them across the panel, just above midway. I use a liquid underglaze or overglaze colour which looks similar to the fired colour, in Germany it is called 'Dekofarbe'. I overlapped the letters and varied the kinds of blue I used to make a more interesting arrangement. Next I applied a strip of clay that I had rolled out thinly. I scored this carefully on the reverse and in the right place on the panel and stuck it in position with slip.

The fine lettering was written freehand with liquid colours, but could have been written with decorating colours in tablet form. I prefer the liquid type as they are easier to control.

The piece was fired in a glaze firing at 1250°C (2282°F). It is 28 cm (11 in.) tall, and is mounted on Plexiglass (29 x 35 cm (11½ x 13¾ in.).

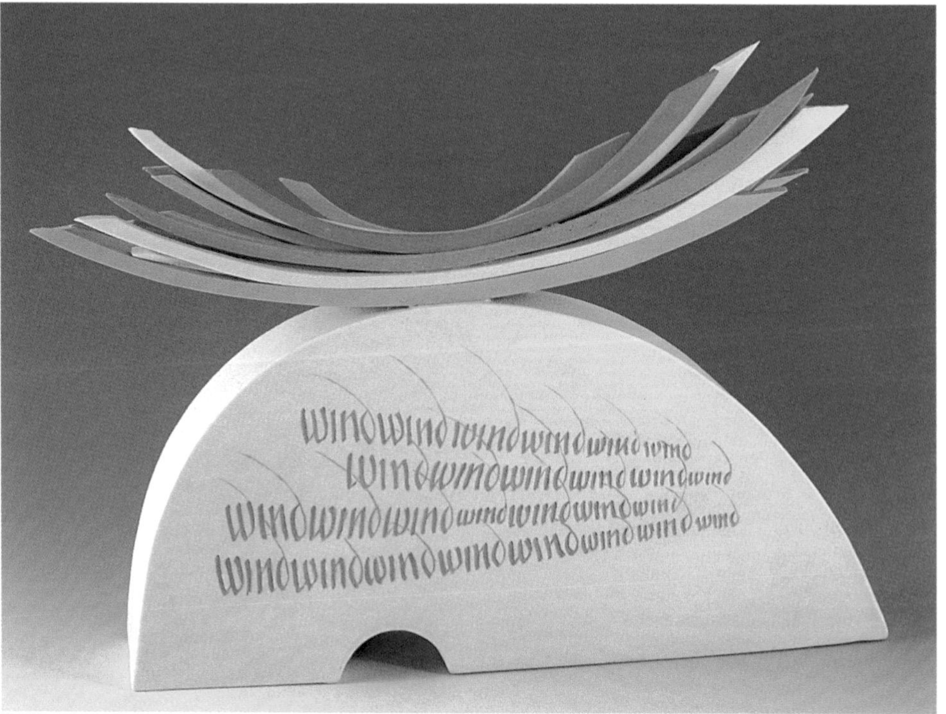

Tiles

Tiles are the easiest forms to make and they are very popular with potters as they are very versatile.

After you have rolled out your clay, cut it into squares the size you want, remembering that clay will shrink when fired, and carefully place them on a smooth board to dry a little. It is a good idea to make a card template of your tile so that you can cut them all the same size. Cut around the template carefully, keeping the knife held at a right angle. If you are not going to work on them straight away cover them carefully with plastic so that they do not dry out, and occasionally turn them over, being careful not to distort them. When they are leatherhard choose the right side and carefully smooth the edges a little if you wish using fingers, sponge or rubber kidney. Study commercial tiles to see how it looks. A series of tiles is a good way to experiment with different methods of using letters and will be a valuable reference for your future work.

OPPOSITE, ABOVE Rocking sculpture, *Chaos* by Mary White. When I made this I was working on half-moon shapes and also the idea of wedges of clay, like slices of melon. This fitted in with my preoccupation with chaos and order in chaos and also with coloured porcelain which I was developing. First I made a wedge shape that was slightly irregular and scratched lines and letters on both sides: these I filled with oxide. I used coloured porcelain to make the cut-out letters and fired them separately. After firing I stuck the letters in random design over the top. The piece can be gently rocked.

OPPOSITE, BELOW *Wind* by Mary White. This was made in a similar way to the rocking sculpture, but it stands on the flat side and the form slopes to a flat edge at the top. It is 32 cm (12½ in.) wide (approx.) I used a lettering nib to scratch the word 'wind' on the sides, trying to convey the idea of movement. I rubbed grey stain in the letters and wiped off the surplus from the surface. At the top I had a problem. I wanted to attach curved strips of porcelain that I had stained with colour to give an effect of movement. It proved too difficult to fire them in place, so I fired all curved strips separately on a clean batt and stuck them in place later. It was an extremely difficult piece to make, but I was pleased with the result. Now the biggest problem is how to transport it to an exhibition! *Photographs by Mary White.*

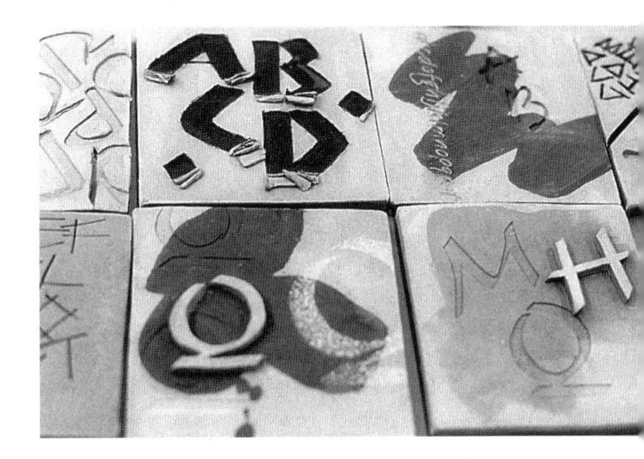

Work by students.

Work by students.

Panels and murals

Tiles can be mounted together to make wall-panels or a whole wall could be tiled. This is major design work as there is a lot to think about in a large work, but it would be a good idea for a school project.

I made a panel composed of 24 tiles arranged in six rows of four. I was experimenting with different ways of using letters on clay at that time and decided to use tiles which would allow me to use different techniques. I wanted to arrange them on a panel, grading the weight of the design from top to bottom and restricting the colour to black and white without a glaze firing. First I drew a simple alphabet in various sizes, simple but with subtle curves and angles that were easy to cut out in card. I used these as patterns to cut out the letters I wanted in thinly rolled clay, they were placed carefully on a board and covered with plastic sheeting. Next I cut out 24 tiles from thicker clay slabs using a metal tile-cutter and placed these on boards, also covering them with plastic. It was important for both letters and tiles to dry flat and to the same state (leatherhard), so I had to control them, and occasionally turn them over. I have found that it helps to slightly tilt the drying boards being careful that the pieces do not slip off.

While the clay was drying I had time to experiment with the card letters to find a way of arranging my design. I did not want to use a glaze, only a black underglaze which I could paint on. Some tiles I decided to leave white as I particularly like the effect of shadows on white clay, and I was using porcelain. These effects I could achieve by impressing some letters and building up the cut-out letters. All-black tiles gave the opportunity to etch around the card letters leaving a delicate line-design. This I repeated in reverse with black lines on white clay. When assembling the letters on the tiles I scored them carefully on the reverse side to help bind them, painted slip on them and placed them firmly in position. Any surplus slip was removed with a soft damp brush. Burrs on the scratched tiles had to be removed when dry. To avoid warping I dried everything slowly in a cool room using plastic sheets.

When the tiles were eventually fired I had to decide how to mount them. I made two sets in order to be able to experiment, so I mounted the first set on a board which I had painted white, leaving a generous margin around them. The second sheet I mounted on a thick sheet of plexiglass, which pleased me more. I drilled a hole in each of the top corners so that I could hang it on the wall with cord or wire. You could alternatively make holes in the corners of your tiles for mounting on the wall with nails or screws, although you need to be very careful not to hit the tile when using the hammer or screwdriver!

Small panels

Panels that will fit onto a kiln shelf can be rolled out at about 7 mm (¼ in.) thick and can be lettered with any of the methods you have learnt, even adding very thin strips or shapes to give more scale and interest. First choose the wording then design the actual size and shape of your panel, remembering that it must fit onto one of the kiln shelves. The edges of the panel need careful thought; whether to cut them accurately, but maybe wipe them with a finger or a sponge so that they are not sharp; to shape them in some way; to add a small strip to make a frame or to leave the panel with

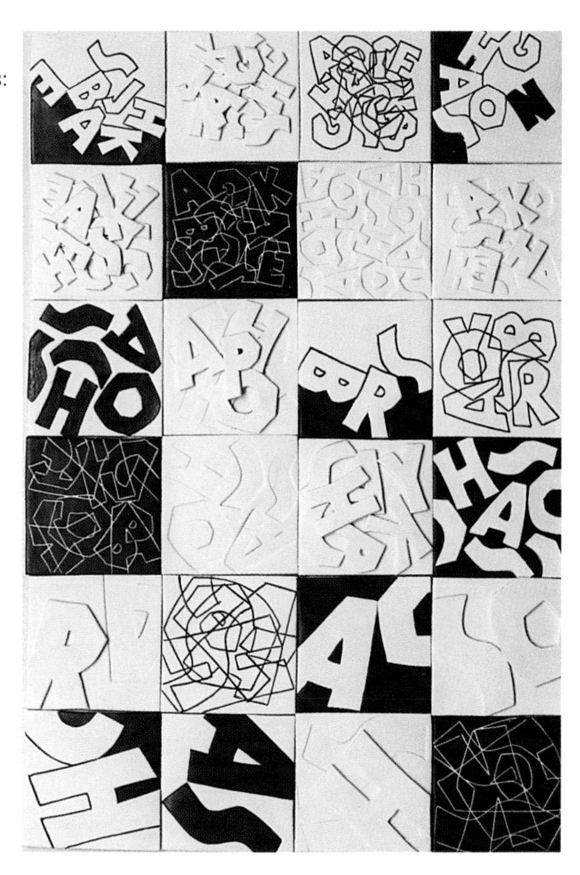

Plaque composed of tiles by Mary White. Mounted on board. Dimensions: 58 x 78 cm (22¾ x 30¾ in.).

rough edges on the natural shape that happened when you rolled out the clay. I discovered a way to get an interesting edge, but it needs practice, by folding the clay over near the edge and then gently tearing it like paper; if you are lucky this forms a kind of deckle-edge. Oxide can be rubbed into the rough edges and then wiped off with a damp sponge so that it only remains in the cracks. A grey underglaze is particularly effective used here.

Coiled forms

One of the oldest ways of making a pot is to coil it. This is a method favoured by many ceramic artists today as it is a versatile and satisfying method and is often introduced in schools. After you have selected and prepared your clay, make a flat base by rolling some clay to the thickness you need for your pot. Cut out a circle using a guide made in paper or thin card and carefully tease the edge to come up a little higher than the base: this is to strengthen the join between base and side wall. Roll out your coils according to the thickness you want for your pot, allowing for flattening the coils which will reduce the thickness. It helps to knead your clay into a sausage shape before rolling. Work on a plaster batt or on an even board large enough to roll the length you want. Cut off a coil long enough to fit on the base just inside the edge so that you can coax the clay up over the coil and seal the join. Seal it

Thrown rounded form with gold calligraphic lettering by Mary White. Width: 14 cm (5½ in.). *Photograph by Inge Mess.*

Thrown bottle form by Mary White, with silver and gold lettering applied. Width: 13 cm (5 in.). *Photograph by Inge Mess.*

inside as well with a thin coil. The base is where the pot could come apart, so take care. Now all you have to do is keep adding coils and join them in rings, sealing them together well to avoid cracking apart. You can use a metal kidney to draw the clay up and to smooth the joins. Inside you can use the curved edge of the kidney to smooth and help curve the shape. You might find a wooden kidney more useful if you have one. Go on adding coils and smoothing them together until you have made the shape and size you want. To form an outward curve add the coils to the outside of the pot; to close it in add the coils to the inside. Always use a coil long enough to go around the coil already fixed and do not let the ends overlap, instead cut them to meet diagonally and join with slip. You can join the coils by pressing them together with your thumb before you smooth with the kidney if you like; whatever you do you must seal the joins otherwise they will open in the firing. Never use a continuous coil as this will distort the shape. When you reach the height you want you must decide how you want to finish the shape. Do you want an open bowl or will you draw the shape in to make a neck? Before you start you should really make sketches and look through ceramic books to see how other people have worked. Read all you can about coiling (try *Coiled Pottery* by Betty Blandino and *Handbuilding* by Michael Hardy, see Bibliography). The important thing, if you want to use letters in some way on your work, is to smooth the surface. Use a kidney and then when it is dry you can gently scrape it or even use sandpaper. Of course any letters you want to cut out and apply must be fixed before the pot is too hard and dry.

For beginners it can be tricky to make a good coil pot and to letter it, but it just takes practice!

Advanced potters, of course, have the great advantage of being able to make any shape by any method, but then they will not have the experience of a calligrapher or letterer!

Cubes

Having tried different techniques on tiles it is interesting to adapt them to a cube form. I find this a favourite shape for students and we usually make them in my workshops as they can be made without any previous knowledge of ceramics. It is also a good shape for design.

First smooth out your clay and cut 6 equal squares (you could use a tile-cutter or knife) and place them on smooth boards to dry a little. At this stage you could start to design your cube, remembering that the design must carry on around the cube and must look good from all angles. As soon as you can work without damage, carefully bevel all the edges to 45°, keeping your knife at the same angle every time. Keep all the triangular strips that you cut off under plastic sheeting with the squares so that nothing dries out. If you want to impress letters you could do it at this stage, but try not to distort the squares, if you do you can very carefully place the tile-cutter over the clay and cut the surplus off. If you have not used a tile-

Porcelain cubes with relief letters, by Mary White. Collection of The Gutenberg Museum, Germany. Dimensions: 10.5 x 10.5 cm (4⅛ x 4⅛ in.). *Photograph by Mary White.*

Box cube with lid and stamp letters, from a student conference in the U.S.A.

Stack of cubes with relief letters on by Mary White. Height 29 cm (11½ in.) In a private collection. *Photograph by Inge Mess.*

cutter you will have to use your skill!

Start to assemble your cube when it is leatherhard by scoring the edges of the squares with a sharp knife or point and make some slip from bits of your clay in a saucer so that it resembles cream. This is used as a kind of glue to bind the edges together and the scoring helps to make a firm join. First build up the four sides around the base and use some of the triangular strips to seal inside the corners using slip and a moist brush. Next fix four strips along the inside top edges to support the lid, using the same procedure. Check that the sides are all regular by placing a right-angled set-square against them and pat it into shape with the flat side of one of your wooden strips if necessary. Remember that clay shrinks as it dries, so make a small hole in the base with a needle, but if you need to press more letters into the surface do not make the hole until you have finished because you could distort the shape. Now you can finish your design, adding letters with the same method of scoring and applying slip, and whatever else you want to do. When you have finished it is advisable to keep the cube under plastic for a day or two so that it can dry slowly, but keep turning it over and make certain that there is a clean surface underneath.

When your cube has eventually dried out you can correct any slight irregularities on the surface by placing it on a clean sheet of the finest sandpaper and gently moving it around. It is also possible to fold the sandpaper around a block of wood and gently rub the surface with this, but be very careful as you can easily damage the work. If letters do fall off you can try sticking them on with powdered clay mixed with vinegar or as a last resort, fire them separately and stick them on when the pot is finished.

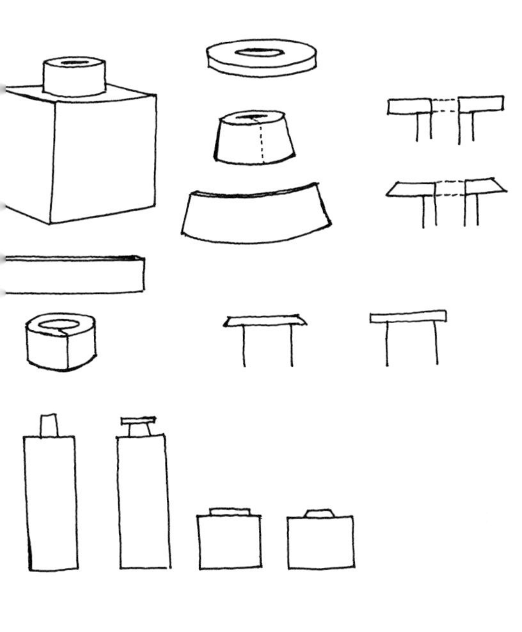

Suggestions for bottle forms from slab.

With ingenuity you can convert your cube into a vase by first cutting a hole in the top slab before attaching it and then adding a neck made from a slab curved around and joined with slip, perhaps with a flat ring on top.

You can also convert your cube into a box, either by making a lid to fit by adding small strips inside the top slab (remember to add a handle also!) or by a very interesting method of cutting through the sides of a finished box just before leatherhard with a needle, curving the cut so that the lid will not fall off. This last method is interesting but makes problems with placing your lettering; it needs very careful and imaginative design.

It is not necessary to make your shapes true cubes. You can vary the proportion so that you have a tall form or a squat one. Another idea is to cut the letters right through the design and adapt it to a lamp base, allowing for the bulb to fit inside, shining through the letters. Once you start designing you will find that you get many ideas, and all without a potters wheel!

Pebble pots

A form that I often use in workshops is a pebble shape because students can work with this method without difficulty and learn something about ceramic processes at the same time. I actually started making them in the 1960s when we lived in Wales near the sea and I was teaching in Atlantic College. The students collected rounded pebbles from the beach. If you do not live near the sea you might find pebbles by a river, but most garden centres have a good selection; look for interesting rounded shapes. Choose several as they look best in a group (and are often sought after for flower arrangements).

Roll out your clay, thickness and size according to the size of the pebble, place it over the stone and gently mould it into shape taking care not to mark the surface with your fingers, then cut around it where the stone is widest making sure the clay is smooth and the cut is even and clean. After a while, gently remove the clay and place it cut side down on a wooden batt keeping the rounded shape, then cover with plastic. Turn the pebble over and form the other side and remove it in the same way. The shapes must be left until they are leatherhard and if they lose their shape gently fit them over the pebble again for a few seconds or minutes, but do not leave them there because the clay shrinks as it dries and the shapes will split. If you want to impress letters it is best to do this while the halves are on the pebble as pressing the final shape would distort it, painting or stencilling can be done when the

Moulding the clay around the pebble. *Photograph by Inge Mess.*

Scoring and putting slip onto the edge of the pebble pot. *Photograph by Inge Mess.*

Joining the two halves of the pot together. *Photograph by Inge Mess.*

The completed pebble pot, standing on its side. *Photograph by Inge Mess.*

shape is finished. To combine the two halves you must score the cut edges and paint them with slip. Then carefully place them together, matching the shapes exactly. You need to work on the join with a flexible steel kidney which is an invaluable tool. The join must be strong and invisible. At this stage you can pat the form with wood to slightly alter or refine the shape if you wish. Decide how it is going to stand and slightly flatten and hollow the base. At the top you can make one or more holes with a cutting tool. On no account leave the form to dry without a hole of some size or it will split. Any irregularities can be scraped off with a flexible kidney. Cut-out letters must be applied with scoring and slip before the

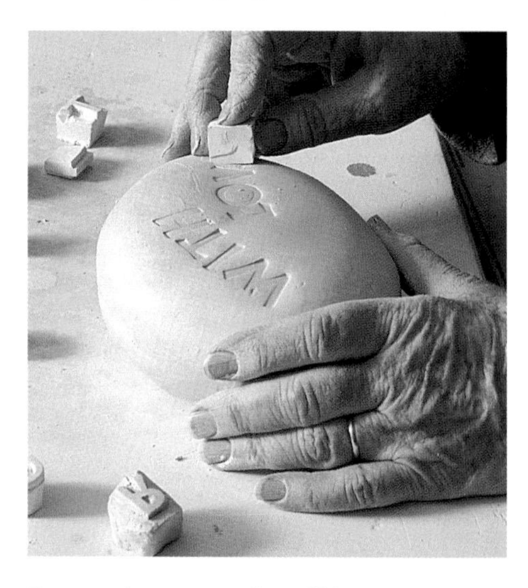

Pressing letters into the pebble pot. *Photograph by Inge Mess.*

clay dries, but letters can be painted leatherhard or dry. Irregularities can be smoothed with fine sandpaper, when dry. When you decide that you only want to paint on your pebble, you can sand it smooth when it is dry enough and then paint with the suitable oxides.

Bowls

If you find a suitably large and smooth stone you can use it to make a bowl. Roll out your clay, not too thin, place it over the most interesting side of the large pebble or stone and cut around the wide part as before. Leave it on the stone until it is beginning to get leatherhard, but be careful it does not split; this will be your bowl. At this stage think about how you could make a foot for it to stand on. The easiest way is to cut a strip of your clay and fold it around an extra-wide candle or similar object that you have already wrapped with a strip of newspaper, then seal the two ends together with scoring and slip. When this is at the same stage of dryness as the bowl (leatherhard) you can slightly modify it so that it fits exactly to the bottom of your bowl. Strengthen the join with small coils and neaten them with wet fingers or a thick damp brush.

With ingenuity you can think of other ways of making a foot, perhaps three pieces of clay moulded into shapes and applied to the bottom.

When it is finished, you can smooth the edge with a sponge at the leather-hard stage or with sandpaper when dry. Letters can be applied, pressed or painted at the appropriate stage. A group of small 'bowls' could be joined together to form a multiple shape. With imagination you could use this method to produce many interesting objects.

Bowl by Mary White, diameter: 28 cm (11 in.), from a private collection, U.S.A.

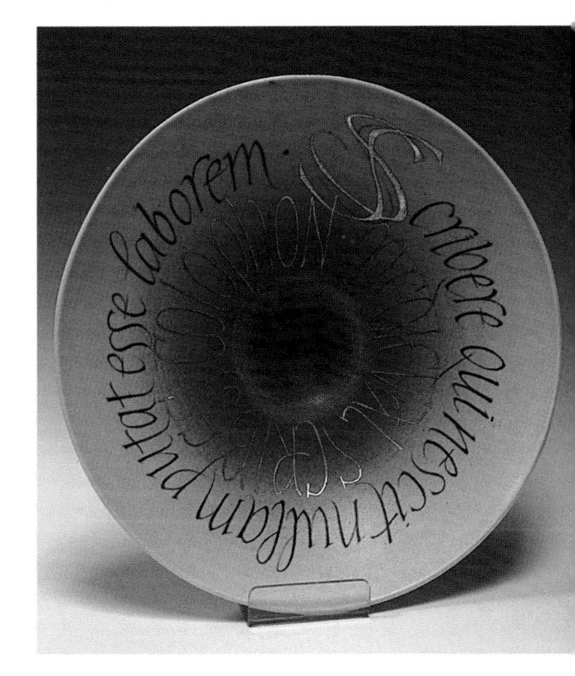

Bowl with lettering and gold lustre by Mary White, diameter: 28 cm (11 in.). Private collection.

Chapter Five
Alphabets

'Lettering' has a wide meaning: it embodies all forms of letters which are arranged together to be read. For instance, type, which is printed; signwriting; letters for monuments and gravestones; graphic design; book production; newspapers and magazines, captions in films and TV. The chief requirement, usually, is that lettering can be read, but this does not mean that it cannot be pleasing and well-designed. A memorial should convey timelessness and dignity; signwriting should be appropriate, not blatant and offensive to the eye; books should be comfortable to read without eye-strain; graphic design and film captions should convey atmosphere and appeal to the right emotions. There is a lot of psychology applied to letters, and skill and training involved in good design. Strangely enough the alphabet that has the greatest influence is one of the oldest – the Roman alphabet, as seen on the Trajan Column in Rome. It is the basis of modern type, even on the computer where it is known as 'Times' because it is used by the *The Times* newspaper due to its legibility and its 'timelessness'. Its characters with thick and thin strokes which terminate with serifs are found in many of the hundreds of type faces in printers' catalogues. Also it is used in various personally adapted forms in stone and wood carving, in signwriting and in graphic design.

Percy Smith, an English lettering artist in the 20th century, designed a new style of letter for the London Underground now known as 'Sans Serif' or 'Block'. This was a revolution! Strokes of equal width and no serifs! It is still used in the Underground and for many street signs as it is easy to read and can be seen well from a distance. Together with the monumental Roman, Block dominates modern letter design in updated forms. These styles can be used on clay and treated in various imaginative ways.

'Calligraphy' is more difficult to define. It can also be called lettering but it is an expressive art as opposed to writing, which should be legible. Calligraphy must not be called beautiful writing ... to a calligrapher that is offensive, it is like saying that a painting is pretty! In fact it is not even necessary to use actual letters: it can consist of sensitive and skilful strokes. A drawing can also be calligraphic. Look at the works of Matisse or Picasso, they were masters of line. This does not mean that the history of lettering need not be studied or that calligraphy cannot be read. It is important for students to be familiar with the history of the alphabets and of old manuscripts and to be able to adapt historic styles to modern use. 'Adapt' is a key word. Much can and should be learned from the past, but to laboriously copy it exactly and continuously is stifling. A painter does not continue to paint like the Old Masters or even Picasso, but he can gain much by studying their work. Contemporary calligraphy is akin to Fine Art: it should convey feelings, even excitement. A skilful vigorous curve can convey as much as a passage of music, in fact it has often been compared to music or dancing. Freedom, rhythm and sensitivity make letters dance.

ABCD
EFGHI
JKLM
NOPQ
RSTUV
WXYZ

Roman capitals based on the Trajan Column, Rome. These have had great influence on lettering and modern typefaces such as Bembo, Times Roman, etc. and on monumental lettering.

half uncials

abcddef
ghijklm
nopqrst
uvwxyz

hmuwy

Half Uncials were first used by the Romans for less important books, early in the 6th century and were brought to Ireland by Roman missionaries, and then to England. In Ireland they were used in the famous *Book of Kells* now in Trinity College Library, Dublin. They were taken to Holy Island off the Northumberland coast and were used for the Lindisfarne Gospels written in the Abbey which was founded in AD 635.

Uncials

ABCDEFG
HIJKLMN
OPQRSTU
VWXYZ&
KNVWXY

These capitals are based on Roman Uncials which were in use from the 5th – 8th century. Today many variations are popular with international calligraphers.

Both uncials and half-uncials were written with the edge of the pen held parallel to the horizontal ruling lines, giving thick vertical strokes and thin horizontal ones. Today there are many variations of this style.

Based on Roman capitals, versals were used in early manuscripts for initial letters and headings. They are often adapted today.

minuscules
based on 10th Century
Winchester hand.

abcdefgh
ijklmnopq
qrrstuvw
xyz&x

æflft raree ryrt?!

Known as the 'Foundational Hand' this alphabet has influenced modern type design. It is the first hand most calligraphy students are taught today.

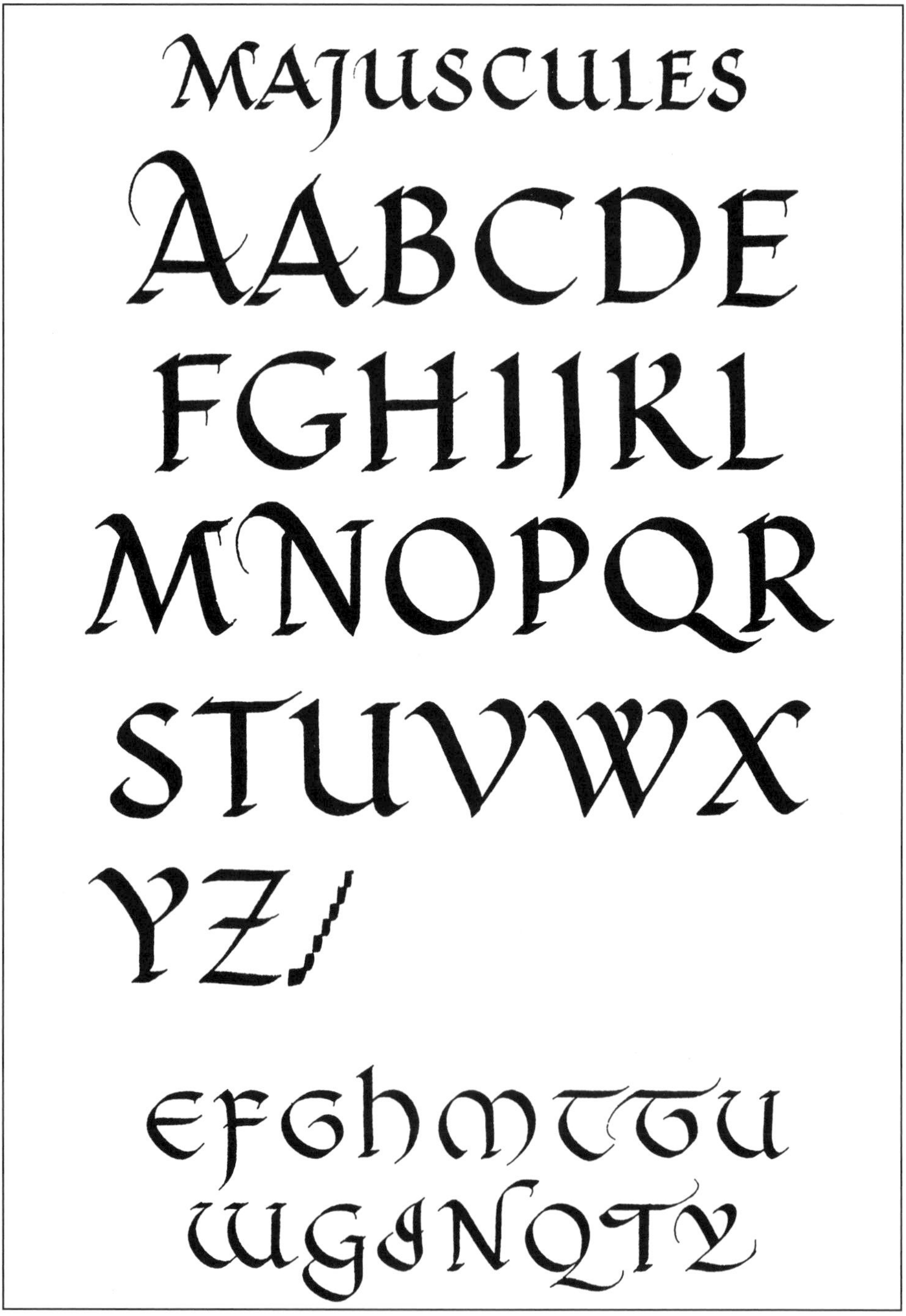

Majuscules

AABCDE
FGHIJKL
MNOPQR
STUVWX
YZ

efghmttu
wgINQTY

Capitals for use with Roundhand Miniscules. Alternatively flat serifs can be used at the top of letters, but this must be consistent.

There are many versions of Gothic which are complicated and very decorative. This is a simplified version of a 13th-century alphabet.

abcdefghijklmn

opqrstuvwxyz

ABCDEFG

HIJKLMN

OPQRSTU

VWXYZ

Italic, based on 16th-century Italian cursive.

Block letters. They were developed by Eric Gill in the 20th century, based on Roman proportions but with strokes of even thickness without serifs. Many variations are used in modern type and graphics, the best known type is Gill Sans which I have used here.

Free forms, an example of how calligraphy is used today.

Chapter Six
Artists using lettering

Australia

Ross Barry

Ross has been involved with the art world since she began studying as a graphic artist in 1959, at the age of 16. She worked in this field for some time in Perth and Sydney and then immersed herself in painting for seven or eight years. Returning to Perth in 1980, she did a four-year course in ceramics and this became a major part of her life.

As a painter she was greatly interested in buildings, so it was natural for her to use architecture in her ceramics. It was a challenge to her imaginative nature to construct her new ideas.

A number of years later she was introduced to calligraphy by a friend and she took lessons, eventually incorporating letters into her ceramics. She enjoys experimenting, and now has painting, ceramics, graphic design, and calligraphy to work with, involving many techniques. As an extensive traveller with her husband and children she has absorbed many textures, colours, environments, histories and cultures that she uses in her work. About half of her ceramics include letters. Her work is rich in design as she spends a lot of time drawing and working out her ideas before she starts. Ross works on leatherhard clay allowing drying time between designs which is speeded up by using a hair dryer. Her calligraphy is integrated into the design. Brush lettering she finds easiest to control on a soft surface. Stencilling she finds very useful, especially if used with an airbrush, as the finish can be varied from soft to sharp, and a sea sponge dabbed over the surface is effective. Often she textures or incises the clay before decoration to give an extra dimension. The finished tiles are dried on a rack for seven to ten days, turning occasionally, and covered with a cotton cloth, weighed down with small weights.

Monument Cafe Slab and mould construction. Roman lettering. Balconies and grills made from metal and glued into place during final construction, then pegged with nails onto the back. The ceramic can be attached to a wall. Inspired by a visit to Rome, sitting in a café amongst pieces of ancient walls and columns.

ABOVE LEFT AND RIGHT *Amplexus Romanorum* (Embracing Romans) Slab construction and moulded roofs. Versal lettering. Inspired by a visit to Roman ruins. Gates and balcony made of metal and glued in place. Freestanding.

From the Runes Strong, heavily grogged stoneware. Runic alphabet and versal alphabet. Perlite was wedged into clay to give a volcanic texture. Glazed with a turquoise/ copper glaze (sprayed on) fired to 1130°C (2066°F).

Cubed House Slab and mould construction. Stamped Roman and personal lettering. Stamped letters and cut-out letters sprayed through with underglaze. Front of building attached to backpiece with pinned nails. Freestanding.

Majolica White Base Glaze

1060°C-1120°C (1940°F-2048°F)

Frit 3302, KGF 9146	90
BBR clay	10
USA bentonite	3

Stains Glaze

Frit KGF 4108	50
Stains	50
USA bentonite	6

Must be mixed and sieved very finely a number of times to a smooth consistency.

Corner Cafe Same technique as *Monument Cafe* and *Cubed House*. Textured sponge and brushed-on glaze on greenware. Fired to 1130°C (2066°F).

Belgium

Vincent Kempenaers

Vincent is concerned with the intolerance in the world and the resulting terror – the Taliban, Klu Klux Klan, etc. – and his ceramics express his feelings about this and also the innocence of children caught up in it. He works with slabs – porcelain when he wants to engrave and stoneware for larger garden pots. When the shapes are large he makes the slabs on the ground and then forms them to stand up. Some are engraved with children's drawings and his personal writing, some are topped with figures made from extruded coils. Sometimes he applies different dry barium glazes in layers and then covers them with a wash of oxides, sometimes he applies a black wash of cobalt, manganese, and iron direct onto the clay. He raw fires once at 1280°C (2336°F) but does a second firing at 980°C (1796°F) for the yellow glaze used for some writing.

Traces 3 Slab-built porcelain. Personalised engraved lettering. Oxides thickly applied. Once-fired to 1280°C (2336°F). Inspired by children's drawings and automatic writing.

LEFT *Garden of Pots* Slab-built stoneware, constructed on the ground. Personalised 'tagged' writing in yellow glaze. Layers of barium glazes covered with black oxide as base for writing. Fired to 1280°C (2336°F) in electric kiln. Second firing at 980°C (1796°F) for yellow glaze.

RIGHT *Intolerance* Slab-built stoneware to be constructed on the ground. Personalised and 'tagged' writing. Dry barium glazes covered with a black wash. Fired to 1280°C (2336°F) then to 980°C (1796°F) for the yellow lettering.

Traces 3 Detail.

Denmark

Bo Kristiansen

Bo Kristiansen of Denmark was a ceramic artist who died young. He was one of the most distinguished and talented ceramic artists of the last century. After being trained in the School of Decorative Art in Copenhagen, Kristiansen set up his first workshop in Bornholm in 1968 but later moved to Copenhagen where he worked for six years as a freelance artist for the Royal Copenhagen Manufactory. He exhibited widely in Scandinavia, Germany, Holland, France, England, the USA and Japan. He was awarded state grants and his work is in many private and permanent collections. He was a member of the Advisory Group for Arts, Crafts and Design of the Danish State Art Foundation (1977–1980), a member of the Danish Craft Council (1982–84) and a member of the International Academy of Ceramics, Geneva.

Kristiansen used a stoneware clay he developed which contained china clay, feldspar, flint and a special clay from Bornholm. His forms were simple and geometric, either thrown or slab built, and his smooth clay gave him the right kind of surface for his intricate letter designs. He used plain letter forms, usually a kind of block with squared serifs and he cut these in outline, freehand, with a sharp knife into the wet clay. Often he used two sizes of overlapping characters: large letters covered by two rows of smaller letters. These were arranged meticulously around the shapes, resulting in intricate sections which he painted with coloured slips occasionally using up to 12 different colours. Sometimes he would overlap

Lettered forms with some gold (using a cold technique). *Photograph courtesy of Museum Boijmans Van Beuningen, Rotterdam.*

an extra large letter or Chinese character. All blended into a pleasing, intricate design. After drying, the entire surface was polished and the bisque fired to 950°C (1742°F). The interiors were glazed and the pieces fired neutrally in a gas oven to approximately 1320°C (2408°F). Finally he highlighted some letters with a cold gold-leaf technique.

The exquisite, immaculate work of Bo Kristiansen is unforgettable once you have seen it.

France

Colette Biquand

Colette states that she works simply in clay without many technical aids. Her work is painted with engobes and sometimes polished. She writes on the sections separately or on the assembled pieces. The first firing is made for solidity, the second for the effects of the fire. If the piece is large she constructs a simple kiln for the firing. Past time is so important for Colette that she writes it into her work. She becomes very involved with this and finds herself pushed into unexpected directions. The sudden breaking of the clay resulting from the heat of the fire evokes the idea of cracking, afterwards destruction, such as one sees in old stonework constructions where you can put your own grafitti on top of grafitti from

RIGHT ABOVE *My Rossette.* Dimensions: (approx.) 80 x 60 cm (31½ x 23½ in.).

Grogged clay stretched thin. Writing influenced by Greek and mathematic symbols, imprinted. Colette calls this work a 'mural' which can be hung like a painting. Inspired by *Pierre de Rosette* by Marie de Londres. The theme is also about fracture, breakage, tearing, with 'crackled' writing as outlined in Roland Barthes' *Variations in Writing.*

RIGHT *Palimpreste.* Dimensions: (approx.) 80 x 60 cm (31½ x 23½ in.).

Similar techniques to those used in *My Rossette.* White grogged clay giving a granular support which also polishes well. A gesture of writing or printing with non-identified signs. Imprints and brushwork.

other periods. Colette wants to show the subtle changes made by time. During the course of history different layers of writing hide one another, with the older ones appearing through cracks caused by time. Since the beginning of time people have left their marks in creating a historical clay record. Colette hopes to add to this through her own work.

RIGHT *Banniere* Dimensions: 25 x 90 cm (9¾ x 35½ in.). Red chamotte clay fired at 1000°C (1832°F) and smoked. Engobes and oxides. Signs made freely with a brush which 'caresses' the clay. Inspired by Chinese fluttering banners.

BELOW, LEFT AND RIGHT Details of *Banniere*.

France

Florence Bruyas

Florence says that her work is like a book that she writes and reads at the same time. She reflects on memories, and souvenirs, and their fragility. Like the archaeologist who treasures fragments of history before they are forgotten, she dusts off her souvenirs and offers them for viewing with nostalgia. She writes the history with which she has been endowed, looking back to her childhood. The earth has a memory; she relives hers, but the memories are not sharp, they are fluid. More than anecdotes they stir tactile emotions, traces, fragments, impressions. Her work plays with what is said and that which is left unsaid. The sense and non-sense of throwing a discreet veil over an individual story.

Florence fills the white paper, engraving the clay with signs that are almost cabalistic, graphic, which bring her work always to the border of painting.

I am continually amazed at the skill with which the French ceramicists can produce slabs of impressive area that remain completely flat after firing.

Moi, plaque. Dimensions: 45 x 45 cm (17¾ x 17¾ in.). Slab-built, grogged clay. Personalised historic styles of writing. Engobes, overpainted with acrylics. An autobiographic work about memory – Florence seeking her place in the family history which influences her character. *Photograph by Yves Simon, Art et Photo, St Etienne.*

Petite Fleur de P'tit Pois. Dimensions: 75 x 75 cm (29½ x 29½ in.). Paper clay, fine like paper. Typographical lettering. Engobes. Plaque form. An autobiographical work with reference to her childhood. She shows her souvenirs on parts of walls. Like the title of a chapter, these words refer to childhood souvenirs where only she can turn the pages. *Photograph by Yves Simon, Art et Photo, St Etienne.*

Jars by Florence Bruyas. Height: 50 cm–60 cm (18–23 in.). Fired to 1000°C (1832°F). Slab-built with refractory, grogged clay with good resistance to thermal shock. Decorated with engobes. Influenced by jars of the Qumran cave. *Photograph by Yves Simon, Art et Photo, St Etienne.*

France

Christina Guwang

Christina makes small sculptures in clay and has exhibited with notable success throughout France, mainly with artists. She is fascinated by ancient cuneiform writing and has developed her own 'curviforme' alphabet with patience and skill. These letters are impressed with precision on her sculptures, which are sometimes abstract forms, sometimes wall hangings. In contrast to these, Christina also makes large 'pebble' forms which are meant to be held in the hands and caressed to feel the textures or, alternatively, the polished surface. The signs fascinate the beholder and encourage them to try to understand the meaning. Christina's wall plaques consist of thinly rolled clay, perhaps with torn edges, sometimes laced together, combined with branches from hedges or old weathered bits of wood found when walking. The texts are often from Chinese poems, translated by Claudé Roy, about feeling the effects of natural wonders, the state of love and simple histories. These texts can be

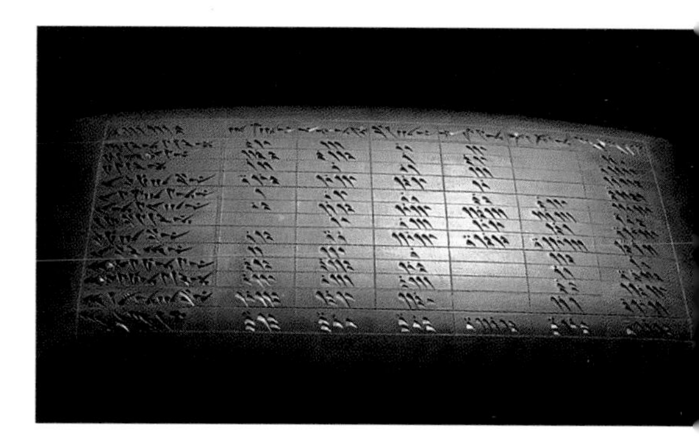

Inventaire. Dimensions: 28 x 14 cm (11 x 5½ in.), fired at 900°C (1652°F).

Black stoneware clay from La Borne with fine chamotte to cut down on shrinkage and limit deformation. Form cut directly from the slightly hardened raw clay, then slipped and polished before engraving the text. Curvilinear development of a personal cuneiform alphabet.

Inspired by accounting records of an ancient civilisation. There is no 0 in the Arcon numerical system, just nine digits composed from three different signs.

RIGHT *Lune d'Hiver*. Dimensions: 38 x 26 cm (15 x 10¼ in.), fired at 900°C (1652°F). 'Curviforme' letters of a Utopian civilisation, the Arcans.

Black clay from La Borne used pure to get the effects of algae in the clay. Christina calls it a 'savage plaque' as she uses these natural accidents. Broken during the firing, and smoked more heavily on one side. The firing accident created an interesting piece with contrasting colours, and it was a miracle that the piece did not explode in the kiln. By sheer luck the beautiful but broken plaque seemed to mirror perfectly the words of an exquisite Chinese poem translated by Claude Roy.

found on her work, encouraging the beholder to try to learn the alphabet in which they are inscribed, or to wonder at them and absorb the mystery. The polished forms are stained with slips, engraved, dried slowly and then wood fired, introducing a chance element: the smoke which can alter the colour.

Inviter la Lune. Dimensions: 27 x 23 cm (10⅝ x 9 in.), fired at 900°C (1652°F). Black clay from La Borne. Again not wedged so that she could use the natural impurities. The white streak in the middle was made with porcelain and ball clay slip and smoked with sawdust. No slip on the base, but lightly smoked. 'Curviforme' letters.

BELOW *Chanson de Tisserande*. Dimensions: 39 x 39 cm (15⅜ x 15⅜ in.), fired at 900°C (1652°F). Curvilinear writing: Arcan. Raku clay (stoneware clay with coarse grog) from La Borne. Slipped before engraving the text. Contrast between the impressed textile and the precise masking-taped edges. Smoking effects reworked with a blow-torch.Text in two stanzas. Chinese poem translated by Claude Roy. Plaques mounted on old drover's sticks used for herding cattle.

Paroles Indiennes. Dimensions: 71 x 30 cm (28 x 11¾ in.). Curvilinear writing: Arcan. Unpugged black La Borne stoneware clay fired at 900°C (1652°F). The orange flashing on the bisque-fired body is due to the action of algae. Smoked with sawdust, no slips.

Rough clay plaques mounted on a combination of driftwood and salvaged timber to create a piece in harmony with an Indian text. Can be dismantled so that all the plaques can be read.

France

Agathe Larpent

Agathe combines paper she has made herself with clay. This is not calligraphy on ceramics but the abstract idea of 'writing' on different media. She puts different elements together to form a plaque or a page in which she creates a vast surface like a window on a journey. She calls it a personal 'murmur' on paper and writes with glaze, crackle, and trailed spots on the clay. A book is 'taken in the hand like a bowl – we touch it, we open it, look at it, read it, close it. We hold a bowl, touch it, feel the quality of the glaze, look at it, enjoy it and put it down.' She makes a history open like a book with a play of details universal in their identification.

Agathe uses glaze, clay and paper as if she is writing with it. She marries techniques of ceramic and engraving in a plastic form. Single pages torn and cracked, are mounted together to form a wall hanging, or hand-made paper is folded and joined to porcelain covers with thongs. All her work is made in porcelain, high-fired and delicately coloured with smoky glazes.

Lieux Reciproques. Dimensions: 65 cm x 75 cm (25½ x 29½ in.). 16 porcelain plaques and self-made paper. 'Just a touch of glaze and a touch of writing'. High-temperature firing. Not really calligraphy on ceramics, but 'writing' on different mediums. *Photograph by F. Xavier Emery.*

Diptyque de Porcelaine. White slab of porcelain and white sheet of paper. 'Murmur' of writing. High-temperature firing. *Photograph by F. Xavier Emery.*

BELOW Similar techniques and materials to those used in *Diptyque de Porcelaine. Photograph by F. Xavier Emery.*

France

Marianne Requena

'Photographs, imprints, and in general all traces, are evidence of existence. Words engraved in the clay have no sense, the history in the details is not important, the essential thing is that they testify having lived.'

RIGHT, TOP Untitled (detail of 90 x 60cm/ 35½ x 23½ in.). Italic writing. Grogged clay necessary for large slab work. (approximately 1 cm/ ⅜ in. thick). Electric kiln fired to 1000°C (1832°F), lightly smoked. Photo-serigraphy-engraved writing. Transparent glaze sprayed on. Inspiration – a familiar place in the countryside.

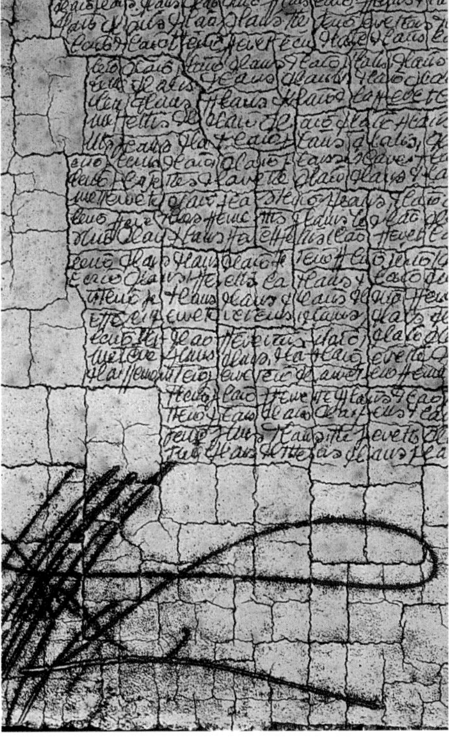

RIGHT, BELOW Untitled (detail) 33 cm (13 in.) x 100 cm (39½ in.). Italic writing. Grogged clay to minimise the shrinkage. Made on a stainless steel frame which allows the creation of large sizes without the necessity of a great thickness of clay. The letters were engraved and then 'inked', as with copper engraving, and the excess wiped off. The inspiration was a trace of life's experience.

Childhood Italic writing. Slabs approximately 1cm/⅜ in. thick. Photo-serigraphy. Engraved writing and 'inked' with black oxide.

France

Sylvie Ruse-Maillard

The ceramics of Sylvie Ruse-Maillard are like poetry. They have the simplicity and spirit of Japanese thinking, but it is a personal art reflecting her life, her beliefs, her knowledge. She lives and works in Chambery, a beautiful, historic town in France, well-known for its old cheese market, a town with small streets, old houses, and Italian influence. Sylvie's house is also her studio and her ceramics are displayed in the entrance. The surrounding mountains are reminiscent of Japanese landscape. Sylvie's ceramics have a quality that makes you want to touch them, to caress the surfaces. Her philosophy is eastern, the aesthetics are Oriental – simplicity, fragments of words and textures – an art of meditation like Zen. She says she tries to work more and more personally. She uses slabs of clay and her inspiration comes from the clouds, the sky and the mountains around her.

Sylvie has studied with several distinguished ceramicists since she started in 1972 and has exhibited in France with colleagues. She holds an annual exhibition in her studio with invited ceramicists and painters.

The Melody of Clouds Dimensions: 9 x 9 x 20.3 cm (3½ x 3½ x 8 in.). Fired at 970°C (1778°F). Slab-built red clay. Personal script. Engobe decoration. Inspired by the mountain sky.

OPPOSITE, TOP *Interior/Exterior.* Dimensions: 48.4 x 39.5 x 0.7 cm (19⅛ x 15½ x ¼ in.). Fired at 970°C (1778°F). Methods and materials as for *Melody of Clouds*. Personal script.

OPPOSITE, BOTTOM *Small Letter*. Dimensions: 26.2 x 20.2 x 7.4 cm (10⅜ x 8 x 2¾ in.). Fired at 970°C (1778°F). Red clay. Slab-built. Decorated with engobe.

France

Daniela Schlagenhauf

After she opened her ceramic studio at l'Escoufoueyre, Daniela Schlagenhauf regularly produced tableware – tea services, vases, bowls, at the same time making bas-reliefs, cubes, cushions, integrating pictorial and musical elements, together with signs and alphabets. The signs appear for themselves and their meaning is obliterated. Using several types of writing of varying age and locality simultaneously, equivalent to a succession of many independent musical themes, constitutes a kind of polyphony of writings of the earth.

The concentration necessary to write the letters, the variable speed of the execution of the signs in their complexity, and the rapid strokes of the brush, determine the flow of time necessary for their mode of making.

Looking is then invited, to penetrate the intimacy of a creation whose seemingly fleeting existence becomes embedded in the viewer's memory.

Daniela's technique is precise, yet free, the craftsmanship superb, producing works of intellectual beauty. With skill she manages to combine red and black clay and porcelain, chosen for their strength in firing. Oxides are applied to the raw clay and fired to 1180°C (2156°F).

Coussin de Lecture. Dimensions: 46 x 46 x 28 cm (18 x 18 x 11 in.). Fired at 1180°C (2156°F). Red and black clay and porcelain chosen for their mechanical resistance. Two pieces stamped in moulds including different motifs in clay and porcelain, attached before firing and re-engraved. The cushion is an image of repose, enriching the idea of reading an open book.

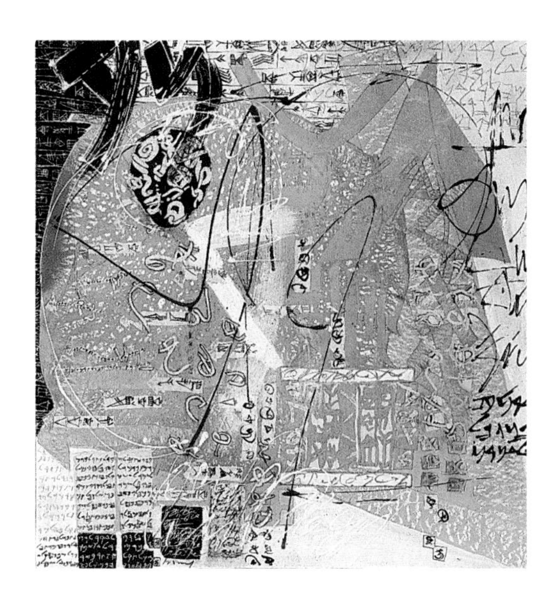

Bas-relief. Dimensions: 69 x 69 cm (27⅛ x 27⅛ in.). Fired at 1180°C (2156°F). Cryptograms of ancient writing and personalised writing. Similar clays as used in *Coussin de Lecture*. Oxides applied before firing.

RIGHT *Triptyque.* Dimensions: 152 x 32 cm (60⅜ x 12½ in.). Fired at 1180°C (2156°F). Similar clays as used in *Coussin de Lecture.* Strips of porcelain were chosen for the central piece for its plasticity. Inclusions of different stonewares and porcelain in a stoneware refractory clay. Oxides applied to raw clay. The ceramic is composed of bas-reliefs conceived as a carpet to be read, linked together by their openwork frame, where the function of porcelain is to provide a material support for the writing strip.

Germany

Rolf Overberg

Rolf Overberg, a German artist whose work I greatly admire, sadly is no longer alive. He was born in Osnabruck and later worked there making both wheel-thrown and hand-built reliefs and small sculptures, becoming more and more interested in the latter. From 1973 he developed book objects, some of which can be seen in the Gutenberg Museum, in Mainz. His clay tablets or slabs are often 50 cm (18½ in.) high, very interesting in design and of executionally high standard. His designs are meticulously designed and constructed but all have a mature understanding and are not stiff. Around 1973 he started to develop book objects by pressing newspaper print into leatherhard clay, in this way contributing to the history of newspaper production. Later his work became more free with letters printed, scratched and stamped then coloured with oxides, including even fantasy letters.

Overberg exhibited in many countries, including New Zealand and Australia and his work can be found in numerous museums and private collections.
He was a member of the International Academy of Ceramics, Geneva and took part in several international symposia.

Unglazed stoneware with chamotte, 1981. Dimensions: 38 cm (15 in.) x 31 cm (12¼ in.) BELOW, LEFT AND RIGHT Details. *Photographs by Inge Mess. Courtesy of Gutenberg Museum, Mainz.*

UK

Maggie Angus Berkowitz

Maggie Angus Berkowitz is not a potter in the normal sense of the word. She calls herself a 'glaze painter'. Mainly she works on unglazed, industrially produced tiles, quarries, wall blanks and porcelain. She uses resist, glazes and oxides and fires around 1070°C (1958°F); some of the glazes she makes herself, some she buys and often alters. Her work involves several firings with varying firing and cooling times.

The finished tiles are installed, under her direction, by professional tilers. Because of the size and cost of her work it is usually commissioned for hospitals, official organisations, swimming pools, etc., but some individual tiles are available – these are often 'bird' tiles used for glaze testing. Each new commission generates a new series of tests – and new birds, usually a very short run. She hopes that one day she will have enough time to make a more useful container for exploration of the limited but unending possibilities of glaze combinations, but for now she finds the birds are just the right shapes.

Sundial. Dimensions: 15 x 15 cm (6 x 6 in.), constructed from tiles. Resists, glazes, oxides on quarry tiles. Inspiration – interwoven, endless, poems by Dorothy Nimmo.

Detail of Sundial.

Her training took Maggie to London, Italy, New York and Japan; she studied with distinguished art teachers, including Kenneth Clark, William Newland and Dora Billington. She had residencies in Japan in 1991, 2000 and 2001 and an exhibition in 2002. Her teaching has taken her to Tanzania and the USA as well as the UK and her work has been represented in many prestigious exhibitions and in many books and periodicals on contemporary and historic ceramics.

Maggie's work is pictorial, but she often uses lettering in her designs. She likes to work in close contact with the organisations who buy her work; she states 'commissioning starts with looking and talking – looking at what has been done in the past, discussing what we might do in the future. Talking about your ideas, plans, dreams, looking at sketches of how I might see them take shape and become real.'

UK

Alan Caiger-Smith

Alan Caiger-Smith is a member of the older generation of British potters and one of the most respected and admired. He has employed and trained a succession of young potters who eventually left to set up their own workshops. His work is mostly wood-fired, tin-glazed earthenware with lustred decoration. He rarely uses conventional letters because he modestly says that he is not a letterer, but his brushwork is strong and always calligraphic in character and immediately recognisable. Some of his earlier commemorative bowls were lettered by a talented scribe, Madeleine Dinkel. Alan's work is usually functional but so beautiful that it is often kept in a cabinet by its owners. Bowls and goblets with masterly lustres are typical of his work. I am the proud owner of one of his 'reject' goblets which I acquired many years ago when I was just starting to pot. Why it was 'reject' I do not know, but he is such a perfectionist that the slightest blemish on his work would cause him to break the pot in the rubbish bin.

Alan states that most inscriptions are straightforward commemorations, which give the maker freedom of design, but it must please the customer, causing some restrictions. Other objects, such as goblets, can be poetic in conception. Bowls and plates can be read at a glance while container forms have to be revolved and read slowly. Lettering must agree with the theme of the words. Not all techniques and glazes are suitable for inscriptions, but Alan finds his tin glaze appropriate for his work.

> For the most part I have made pots to be used and lived with, painted in a variety of colours, including reduced lustre and often with strong calligraphic

Bowl with inscription. 'Hidden lettering' recording a marriage. Dimensions: 33 x 12 cm (13 x 4¾ in.).

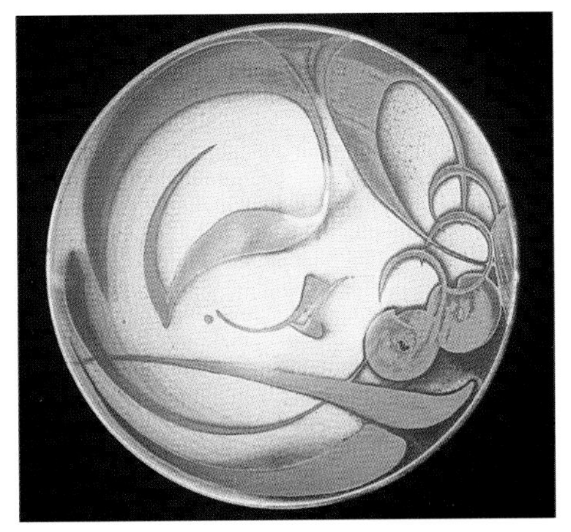

Bowl with brush design. Dimensions: 27 x 8 cm (10⅝ x 3⅛ in.).

brushwork. If some of them are art objects, so much the better. For much of the time I have worked with a team of seven or eight assistants. Therefore some of the designs had to be simple enough to be learnt and repeated. Others are more intuitive and definitely 'one-offs.' I only partially understand these words of T.S. Eliot but they have been an open-ended inspiration to me:

Words, after speech, reach
Into the silence. Only by the form,
  the pattern,
Can words or music reach
The stillness, as a Chinese jar still
Moves perpetually in its stillness.
(*Burnt Norton*)

Alan's best tin glaze for a wide range of colours, firing at 1040°–1066°C (1904°–1951°F).

Lead bisilicate	61
China clay	5
Cornish stone	12
Calcium borate frit	10
Zinc oxide	2
Tin oxide	10

Large bowl with brush design.

The Physic Garden, Ceramic mural. Completed work in situ, dining room, Mary Erskine School, Edinburgh.

UK

Margery Clinton

Margery's origins are in Glasgow and she has seen it change, but it has influenced her. She first studied painting at the Glasgow School of Art and in 1969 she moved to London. She says that delight in the visual world and its elements – colour, light, shape and pattern – have occupied her all her life. During the period in London she discovered the lustre glass of Louis Comfort Tiffany which fascinated her and in 1973 she began research into the technique at the Royal College of Art, and has written a book on the subject. In 1978 she returned to Scotland to set up a pottery in Haddington where she continued to develop reduction lustres and was assisted by Evelyn Corbett for many years. In 1995 she moved to Dunbar and works there alone. She has become increasingly involved with architectural projects with an interest in tiles. Her work can be seen in distinguished collections such as the Tate Gallery, the Victoria and Albert Museum, Glasgow Art Gallery and the Royal Museum of Scotland.

Margery has made ceramic murals for the Mary Erskine School in Edinburgh, the new Glasgow 'Wally Closes' and for the staff toilets at the Scottish National Portrait Gallery in Edinburgh. Ceramics applied to walls within a building are her chief interest, on grand or humble scale. In a public building the decorative ceramic requires very little maintenance and is practical and robust.

The mural for the Mary Erskine School is a project that Margery particularly enjoyed and it is one of the best examples of her work. It was commissioned by the former Pupils' Guild. Although the school had been founded in 1694 the present flat-roofed building dates from 1964 and was without any interior decoration. Margery chose the 16.5 m (54 ft) long wall of the dining room for her mural, which consists of four panels, and it was

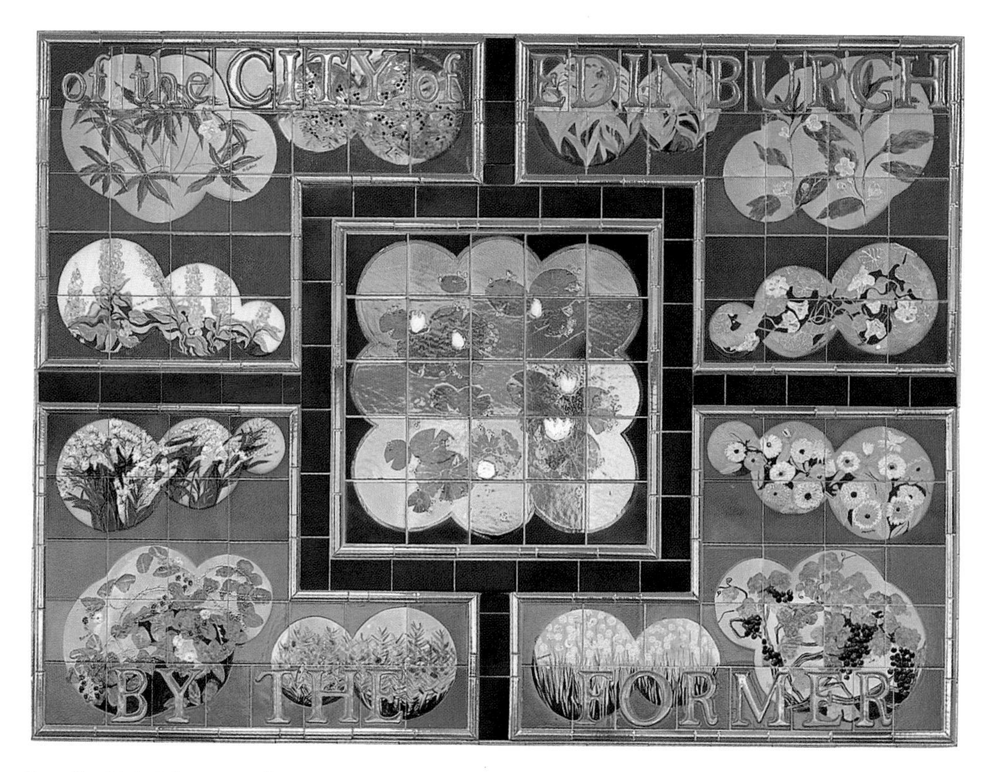

Detail of *The Physic Garden*. One of the six panels of the mural. Dimensions: 15 x 15 cm (6 x 6 in.). Earthenware bisque tiles. Tube-lined lettering against a decorated background. Letters infilled with in-glaze gold reduction lustre. Fired at 1040°–60°C (1904°–40°F) with a brief reduction at about 700°C (1292°F) using her usual reduction lustre technique. She was inspired by Elizabethan lettering where letter forms combine with a decorative background.

presented to the school on 30 June 1994 in the presence of HM the Queen and HRH the Duke of Edinburgh. Margery had done much research into the interesting history of the school. The founder, Mary Erskine was twice widowed and with her second husband, an apothecary, she lived near the physic garden at Trinity Hospital which evolved into the Edinburgh Botanic Garden. A map of Edinburgh drawn in 1647 shows several formal knot gardens. Margery made the mural based on the formal design of a knot garden of that time and filled it with medical plants, many of which would have been sold in the apothecary shop which Mary Erskine continued to run after the death of her husband.

Margery felt that it was more important to fit this 16th-century design into the modern building than to create it in a modern aesthetic. The building is gracefully functional with overhead neutral light, perfect for a mural in reduction lustre as the technique creates a light changing surface on the ceramic. The main pattern is provided by the strong shapes of the formal garden, divided up by a black path edged with lustred tiles with a shallow relief. For the shapes of the individual beds she used intersecting circles, which give a softer secondary pattern which can also be

Macaulay gallery sign. Making details as above.

read from a distance. Within this she designed many medicinal plants featured in the *Hortus Medicus Edinburghensis* of James Sutherland, keeper of the physic garden in 1683.

The commissioners had requested that the mural could be made in a way that it could be removed if it was later necessary. To achieve this the tiles were mounted on marine ply panels, easily removable, which also gives the panels the impression that they are floating on the wall. The mounting and lettering were carried out by Margery's assistant, Evelyn Corbet.

Margery says that despite her age she is open to new developments and some of her latest work is on paper clay, a new technique which has been developed by potters. About 1.5% paper pulp is added to the clay and it totally changes the clay's character. Working with paper clay allows much greater freedom to the artist at the making stage and presents new opportunities.

Earthenware glaze maturing at 1040°C (1904°F)

Potclays leadless 2206	90
Potclays low sol 2204	10
Copper carbonate	1.5
Silver carbonate	0.4
Iron oxide	0.5
Bismuth subnitrate	5
Bentonite	1.5

UK

Laurence McGowan

Laurence McGowan finds the alphabet endlessly seductive! He became interested in lettering and pottery in school but later worked as a cartographic draughtsman. However, the lure of pottery was strong and in 1973 he worked in a pottery in Lincolnshire before he joined Alan Caiger-Smith's workshop where he was able to develop his interest. In 1979 he left to set up a workshop in Wiltshire in the historic town of Marlborough, and started his own series of tableware and commissioned objects and work that he makes for his own enjoyment. I am continually amazed at his skill and patience as he does all his brushwork on the unfired glaze. He acknowledges that this has its drawbacks, but then what technique

Plate. Diameter: (approx.) 33 cm (13 in.). Wheel-thrown buff stoneware. Roman lettering. Brushwork decoration. Biscuit fired to 960°C (1760°F). Cornish stone/dolomite glaze and zirconium silicate. Dip-glazed then fired to 1260°C (2300°F). Inspired by a favourite quotation, of unknown source.

doesn't? He was taught that way while he was in Aldermaston and it has long since become second nature. It takes care and forethought in order to avoid smudging as it is almost impossible to correct any mistakes. He says that 'mind's eye' planning has to be fairly thoroughly thought through, and then the courage summoned-up to be spontaneous and 'fly by the seat of one's pants'! After all, the worst that can happen, 'is that I have to wash the pot off to be reglazed another time, if I can't resolve things or my concentration has slipped'. Laurence uses 3–6mm (⅛ – ¼ in.) inch chisel-shaped brushes along with size 0 for pointed serifs. Because the surfaces are usually curved and because of his method it is impossible to transfer designs from paper, therefore he must know exactly what he wants to do and 'just do it'! He enjoys lettering and studying examples, is fascinated by the abstract patterns they make, and is in awe of lettering as a means of communicating. Lettering on ceramics, especially his method, is a challenge, but he believes the extra expressive dimension it gives a pot is well worth the struggle.

Flat rim dish. Diameter: (approx.) 40 cm (16 in.). Wheel-thrown white 'Earthstone' stoneware. Free Roman lettering. Brushwork decoration. Biscuit fired to 960°C (1760°F), dip-glazed, glaze fired to 1260°C (2300°F).

Alphabet plate. Diameter: (approx.) 33 cm (13 in.). Inspiration was the lure of the alphabet! Technical details the same as for flat-rim dish.

UK/Germany

Mary White

At art school I became interested in lettering and calligraphy and eventually studied at the Hammersmith College of Art (London) because Daisy Alcock, one of the most prominent calligraphers of that time, taught calligraphy and Vera Law was in charge of bookbinding. Then in 1949 I went to Goldsmith's College in London to gain the Art Teacher's Diploma. At the same time Charles, whom I had met at my first art school in Wales, studied at the Slade School of painting, and we became engaged. We returned to Wales, Charles setting up a commercial art studio while I taught in schools, and Atlantic College.

Typography flourished in my department but the students also wanted to make ceramics. At first we brought in a potter to teach them, but later I taught both printing and ceramics until we were able to find a printer to take over the press. It was my my first experience of letters and clay combined! I joined the South Wales Potters and later I was elected to join the Craftsman Potters Association and the British Crafts Centre.

In 1975 we decided to leave Wales and move to England nearer London. We found an old 16th-century pub in Malmesbury and set up studios and a gallery. By now I was producing individual forms, mostly in porcelain. Since this time I have approached ceramics not as a potter, but as an artist.

Then, living in an ancient town with a history of manuscripts, I felt the urge to continue with calligraphy. I made an alphabet of uncial letters in porcelain, like pieces of type, and printed the Malmesbury Charter on both sides of a large coiled construction. The Freemen of the town acquired it to stand in their historic courtroom.

In 1977, the Queen's Silver Jubilee year, I was commissioned to make three pieces with lettering for the Victoria & Albert Musem. It was a great honour but I was not pleased with my results, one piece is in the V & A collection, but I am not at all proud of it as it is not typical of my work. Such is the unpredictable artistic temperament!

In 1980 we decided to move into Europe. We were fortunate to find a former kindergarten in Rheinhessen. I was encouraged by a visit to Professor Zapf, who is the most respected type-designer and letterer in the world. Then I got involved with individual objects in porcelain for international exhibitions and there was not enough time to continue with lettered experiments. In 1982 I was awarded the Staatspreis for Rheinland-Pfalz and was later invited to join the Handwerkskammer as a 'Nebenan' designer member. Villu Toots held a workshop near Bruges in Belgium at this time which was an unforgettable experience. Calligraphy was a new art-form. Gone were the restrictions, and

Thrown form, gold lustre lettering, (painted). In collection of S. Illinois University, U.S.A. (Museum Prize).

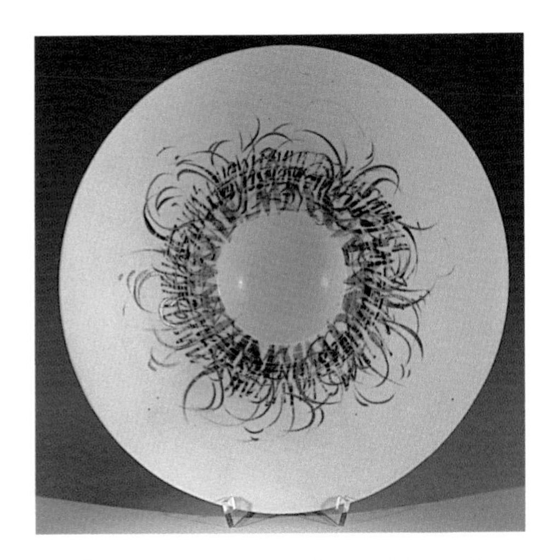

Calligraphic Bowl. Diameter: 25 cm (10 in.). Initial firing 1250°C (2282°F).

Since 1980 I have made a series of wide-flanged bowls in porcelain with calligraphic marks around the central depression. Each bowl is individual, but the idea is basically the same. This bowl is different as there are letters hidden by the calligraphic strokes. The strokes and letters were painted with black underglaze using a flat sable brush. The lettering was painted in black underglaze on the biscuit which was then fired at 780°C (1436°F).

work was imaginative, free and inspiring. Lettering could now be used as a vehicle for self-expression from the heart, just as well as painting or writing. I have always tried to express my feelings in clay, now I wanted to do the same with calligraphy and even combine them.

In 1997 I was beginning to form new ideas when I was invited to teach 'Letters on Clay' at a conference held at the South Illinois University in Illinois, USA. It was a fascinating experience – calligraphy had become an important art in the U.S.A., partly because of the teaching of Donald Jackson and other inspiring calligraphers. My students were mainly graphic artists or teachers and they produced imaginative work. We held an exhibition of our work in collaboration with the University Museum and the *Letter Arts Review* magazine. They awarded a Museum Prize

ABOVE Globe. Diameter: 15 cm (6 in.). Blue and black underglaze on biscuit. Barium glaze fired to 1250°C (2282°F). Gold lustre fired onglaze to (approx.) 780°C (1436°F).

Here I was wanting to acheive layers of interest.

LEFT Globe. Diameter: 15 cm (6 in.). Blue oxide sponged on biscuit. Black underglaze strokes. Glaze fired to 1250°C (2282°F). Gold lustre strokes added in black and fired to (approx.) 780°C (1436°F).

Pyramid with crater. Base (approx.) 25 x 25 x 25 cm (10 x 10 x 10 in.)

The porcelain craters were made with porcelain that I coloured myself. It was a complicated construction. I broke the outside walls, lettered them and stuck them back in place with slip. The piece was fired to 1250°C (2282°F) and the gold lustre later to 780°C (1436°F). Once finished I neatened the edges with a mixture of plaster of Paris and rubber solution (my invention!). I fixed a piece of coloured mirror at the bottom of the crater.

RIGHT Boxed Crater. 25 x 25 x 8 cm (10 x 10 x 3 in.). The inner layers are in unglazed coloured porcelain except for the white one which is lettered with black underglaze. The outer layer has a natural texture made by impressing. A piece of gold mirror in the centre of the crater adds interest and can be seen gleaming from a distance.

which I was fortunate to win, so this ceramic is now in the University Museum.

I had a very academic training in calligraphy and design in the late 1940s, and it was not until the 1960s that I started to use clay in my art work. When I introduced lettering to my objects it was difficult to break away from my disciplined training and I mostly used traditional lettering and calligraphy. Now I am more interested in calligraphic strokes.

When I started to work in clay I used stoneware, but I soon discovered that porcelain gave me the opportunity to experiment with coloured glazes and also to colour the clay itself with oxides. Later, when I started to use letters on clay I was strongly influenced by traditional calligraphic colours with the addition of gold. Mostly I painted the letters with a fine brush and it was much like a continuation of my technique on paper or vellum. I found there is a lot to learn about using letters on clay so I explored as many different methods as possible.

Colour has always been important to me. I learnt a lot from fabrics as I have always made many of my own clothes, and from weaving which made me excited by the effects obtained by mixing coloured threads. Also I am a passionate gardener and plants produce amazing colour combinations.

Since my husband died 7 years ago I have spent more time thinking, gardening, writing and teaching. Now my ideas are more abstract and when I resume my artwork I think it will have developed in a different way. I am fascinated by working in layers and hope to continue this in both calligraphy and ceramics, building-up letters, which perhaps will make them difficult to read but will create the kind of atmosphere I feel. I do not think age should hinder imagination as long as one has the ability to go on working.

UK

Mary Wondrausch

Mary Wondrausch is an artist potter living and working in an 18th-century stable in a 'magical setting'. She specialises in individual commissioned commemorative plates made in a fascinating traditional way. This is slipware – a very lively, sometimes humorous style of pottery that has its origins in earlier times. She combines her skills in the physical making of large cheese platters with painting in the centre and lettering on the rims. Sometimes she uses a more painterly style with sgraffito fish and bird motifs using slips and oxides. She says her work is functional rather than sculptural. Mary has written an absorbing book on the whole subject (*Mary Wondrausch on Slipware*). It is a method that looks simple, but is extremely difficult to make.

Punch and Toby, 'That's the way to do it'. Diameter: 40 cm (16 in.). Fired at 1070°C (1958°F). A wheel-thrown mixture of Fremington and Spencroft clay. Slip-trailed with honey glaze over. Inspired by Thomas Toft, 1667.

Two individually commissioned plates, after Thomas Toft. Wheel-thrown Fremington plastic clay low fired. Slip-trailed with honey glaze over.

BELOW A cheese platter with 'just writing'. 1070°C (1958°F). Wheel-thrown. A mixture of Fremington and Spencroft clay. Sgraffito with painted slips and oxides. Clear glaze. Inspired by her love of cheese and Matisse!

Holland

Anneke Harting

Holland is small but rich in artists whose work is not often seen outside their country, possibly because it is appreciated there and the artists are kept busy.

Anneke Harting, a creative and talented artist, has been fascinated with clay since her first experience with it in her primary school, when a professional ceramicist demonstrated and allowed her to sit on the stool of his potter's wheel. Later she took several creative courses at the Dutch School of Artists in The Hague and in the Ceramic Academy at the Crabeth College of Gouda, where she now teaches.

Anneke developed her own style during her training. Refusing to accept limits imposed by the material, she works with stoneware because of its rough texture and uses it to make vases that are monumental in conception.

ABOVE (detail) AND BELOW *San People.* Biscuit fired to 1140°C (2084°F), glaze to 1240°C (2264°F). Slab-built stoneware. Decorated with coloured pigments and transparent glaze. Inspired by a visit to see the paintings of the San People in the 'middle of nowhere', guided by a well-known South African historian.

Besides the vase forms she also makes ceramic and bronze sculptures. Abandoning the wheel, she slab builds in clay, creating objects with 'exuberant curls, elegant curved and irregular edges'. The colour can be delicate or strongly earthy. Anneke's inspiration comes from language in its many forms, ancient and modern, particularly Eastern cybernetic characters and prehistoric symbols. She searches for texts from 'long ago and far away'. Haiku is her favourite form of poetry. The resulting ceramics are fascinating objects that reach the heart and can best be described by a quotation that embodies her philosophy:

> 'Work is the great reality
> Beauty is the great aim'.

OPPOSITE *Dance*. Technique as before. Anneke became highly inspired by the form and the meaning of the Chinese characters.

Power of Life is Meditation. Technique as before. Both Chinese and Japanese characters inspire her. Many of these characters appeared in her early work, mostly in blue and green.

For Helenè. Technique as before. Lettering taken from computer program. This was made for one of Holland's most well-known PR companies.

North Africa/France

Sylvian Meschia

Sylvian Meschia was born and spent his childhood in North Africa and the forms, colours and sensations that surrounded his early years are clearly perceptible in his work today. With no formal training, he learnt his skills as a ceramicist with the native craftsmen of Tunisia, then in the workshops of artists from southern France, before establishing his own studio in the rolling countryside south of Toulouse, where he lives and works today. His Mediterranean origins drew him naturally to working with the soft earthenware and warm, ochre shades traditional to the area, and he has forged a distinctive style through his constantly evolving and original decorative techniques.

His early work involved renewing and revisiting a local style: slip clays coloured with cobalt, iron and copper oxides are applied in successive layers to a raw, hand-thrown pot to produce a swirling, marbled effect, reminiscent of the traditional earthenware of Roussillon. At the same time he was developing and perfecting his brushwork, inspired, like many ceramic artists, by Japanese pictorial and calligraphic techniques. This led him initially towards bold, simple brushstroke designs on hand-thrown pieces, then ultimately to working on a flat surface, with increasingly elaborate use of paintbrush, bamboo, imprinting and engraving. The tiles he thus creates are conceived as small ceramic pictures, the culmination of his pictorial work, first on paper, then on clay, applying his potter's techniques. He feels that the austere, flat, square format allows much greater freedom of graphic expression than an already expressive thrown or moulded form. Some are combined to form larger wall panels for integration into architectural designs. The inspiration is again clear – Mediterranean – with the use of white set against contrasting colours and splashes of blue, as well as light and

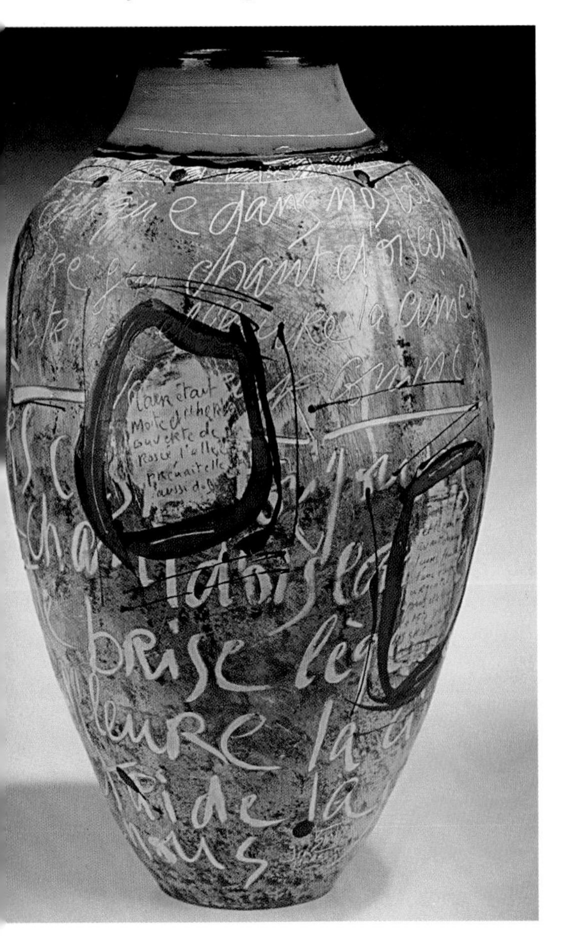

Legends of Autumn. Height: 70 cm (27½ in.). Personal calligraphy. Two local earthenware clays blended to give a deep orange terracotta shade and a highly pliable clay, perfect for throwing large, thin-sided pieces. These pieces are thrown, decorated with successive layers of coloured slip clays, then engraved. Transparent glaze, giving a semi-matt finish, which is then textured with brushstrokes. Inspiration from *Legends of Autumn* by Jim Harrison. *Photograph by Studio Jarlan.*

shade effects created through the slightly textured surface.

Another influence has also been at work throughout the creative process – the artist's long standing love for the written word. An avid reader of poetry and prose, with an innate sense of textuality, Sylvian would frequently note down or memorise snatches of text, culled here and there from his reading, with no other aim than to fix a fleeting evocation, or to savour the sounds and rhythms. These citations have gradually found their way onto the clay, some meaningful, others simply suggestive, always setting up echoes in the ear or eye.

His more formal calligraphic style, freely adapted at the outset from Moorish designs, has evolved to become an individual repertoire of signs and figures, elaborated over time and now entirely his own. Here, from his original two-dimensional engraving, his technique has become increasingly refined and ambitious, culminating in large, hand-thrown pieces of impressive dimensions: free-standing jars, urns and lidded caskets, whose rough, mottled and engraved surfaces seem to speak to us from beyond time.

Renaissance. Height: 70 cm ($27\frac{1}{2}$ in.). Inspiration was taken from a series of short poems by the French poet Michel Houelleberg. *Photograph by Studio Jarlan.*

Que reste-t'il de nos amours? Tile, dimensions: 20 x 20 cm ($7\frac{7}{8}$ x $7\frac{7}{8}$ in.). Inspiration from a song by Charles Trenet. *Photograph by Studio Jarlan.*

Switzerland

Bernadette Baumgartner

Bernadette Baumgartner's work has a mystic quality of meditation and comtemplative simplicity that is intriguing; it is not surprising to learn that as a 16-year old she went to a convent in Brussels to learn French, and thought about becoming a nun. But it was the time of the 'Flower-Power Hippies', when young people became independent, so she went to London to be an au pair instead, and then to Mexico, the USA and to South America. Next she hitch-hiked from Zurich to New Delhi and worked in Europe, Africa and the Karibik as a tourist guide. In 1981 she encountered ceramics and realised she had found her way of expression, and of living, and began to teach herself. Two years later she opened a small ceramic studio. She then trained formally in art schools in Zurich and Bern.

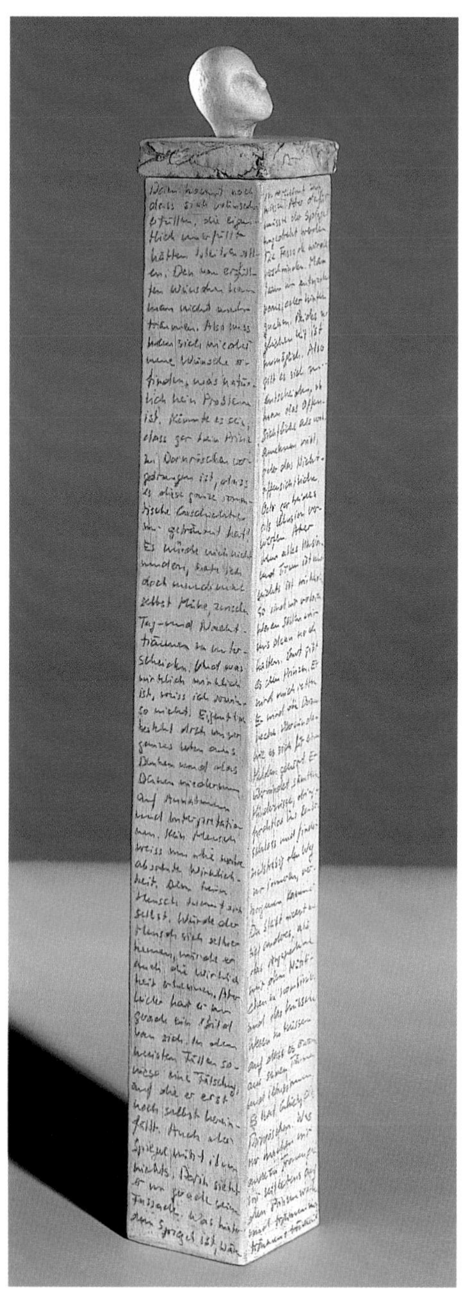

Seelengefässe (soul vessel). Dimensions: 5 x 5 x 50 cm (2 x 2 x 19⅝ in.). *Photograph by Hans Eggermann.*

LEFT Group of *Seelengefässe (soul vessels). Photograph by Herbert Hegner, Au.*

Kleine Kostbarkeiten (*Little Treasure Boxes*). Dimensions: 7 x 7 x 10 cm (2¾ x 2¾ 4 in.). *Photograph by Herbert Hegner, Au.*

In 1992 she opened a new studio in Büron and in 1995 she won an international prize. In 2000 she spent four months in India. On her return to Switzerland she opened a workshop with selling facilities in the Geuensee and has also worked for the Laugenthal porcelain factory since 1994.

Bernadette found her philosophy in India. 'Artha' is the name of her studio; it is Sanskrit and means 'riches and abundance of the earth'. She uses stoneware or a mixture of porcelain and stoneware, building her pieces from slabs assembled, modelled and polished. Her firing takes place in saggars, bins and sometimes in the open fire, playing with chance and risks, and letting herself be surprised by the results. The excitement of the unexpected is part of her work philosophy. She shuns glazes, letting the effects of the fire on the polished surfaces create an unforced beauty, mostly bearing her own text and poetry in a natural way of writing.

The container form is typical of her work – some are quadratic and closed with a lid that may be reversed and might have unexpected letters written inside.

Bernadette has exhibited her work in many galleries, mostly in Switzerland, and in 2002 she designed a small book containing some of her work and her poetic writing which has the quiet fascination, belief and assurance of her ceramics. She is a philosopher in clay.

Vergehen-verwandeln. Dimensions: 25 x 6 x 30 cm (9⅞ x 2⅜ x 11¾ in.). *Photograph by Herbert Hegner, Au.*

USA

Participants in the 'Confluence' conference of 1997

In this section I am including people attending my first workshop in the 'Confluence' conference at the S. Illinois University in 1997. It is interesting to see what can be done in a few days by people who are letter enthusiasts but have little or no experience with clay.

Calligraphy is extremely popular in the USA. Most states have their own societies and each year a well-organised conference is held in one of them, usually in a university compound while the regular students are on holiday. Both the faculty and students are international, and standards are extremely high. Apart from the daily workshops there are various activities, the John Neal bookshop, rooms devoted to papers, inks, writing tools, equipment and exhibitions, etc. Evening lectures and demonstrations are popular and on the last day all participants exhibit their work with healthy competition. In the middle of the week there is a break for a half-day trip to an interesting venue; in Illinois we went on the Mississippi. The food is always good and plenty of it! I can thoroughly recommend these conferences if you have at least a basic understanding of letters and calligraphy. Bookbinding is also part of the curriculum and is modern and fascinating in approach.

Colleen Cavin (USA)

Colleen enjoyed making 'pebble pots' although she was not used to working in clay. Her slides show a pot that she made after going to the conference, using a stone from the river and stoneware clay. She only bisque fired and wrote the lettering on afterwards with gouache. The word 'still' she gilded with 24-carat gold. The wording was from T.S. Eliot: 'Teach us to care, Teach us not to care, Teach us to be still.'

Pebble Pot Cut-out 'Textura' letters which are easy to cut because of the straight lines.
The letters were pressed deeply into the clay to create interesting shadows.

Pebble Pot Roman type letters added with a pen on the greenware.

Sylvia Kowal (USA)

Sylvia sent me slides of pieces that she actually made in the conference. I am not certain what the clay was, apart from knowing it was earthenware.

ABOVE Neuland Box. Cube form. 'Neuland' letters are easy to cut out in clay. The letters were attached with slip after some colour was added and marks made with a dry pen and a needle.

ABOVE Pebble form with impressed letters.

RIGHT Pebble form with letters. Painted.

Marcia Smith (USA)

Marcia began studying pottery in 1971, and has been studying and working in the field ever since, but at the same time teaching intensely, which allows her only limited time for her own work. When she has time she does both handbuilding and throwing, working mostly in porcelain, electric-fired to cone 10. In 1997, she said that her functional work was painted with coloured porcelain slips when leatherhard, then detailed with sgraffito, and fired with a clear glaze.

Since attending the conference in 1997 she has experimented with new ways of using letters. I was pleased to see her work illustrated in the best USA lettering magazine and asked her to contribute to this Gallery.

No 56 x 56 x 10 cm (22 x 22 x 4 in.). Self-cut rubber stamp. Raised letters are simple thick monoline contemporary pointed pen letters. Stoneware A7: a body in which she can make large pieces and fire them many times without trouble. Slab-made and press-moulded with thrown foot.

Letters applied, carved and incised while leatherhard. Black underglaze sprayed when dry. Fired to 1260°C (2300°F) in an electric kiln. Stamped with homemade rubber stamps and stamp pad using underglaze. Script lettering with pen and underglaze.

Clear glaze painted onto 'nix' and 'nil' and fired to 1025°C (1877°F).

The inspiration? One day being asked to do one too many things, leading her to think of all the ways she could say 'no'!

The piece is intended to be wall mounted.

Muse of Fire 61 x 58 x 10 cm (24 x 23 x 4 in.). Slab, press-moulded stoneware, thrown base. Contemporary capitals. Pen lettered caps are based on a Neugebauer letter. Stoneware. Letters cut-out and applied. Incised with pencil when clay reasonably stiff. Sprayed with black underglaze when dry. Fired slowly to 1260°C (2300°F). Clear glaze brushed to all applied letters and to triangle-shaped confetti. Underglazes spattered on with toothbrush. Blue lettering written with pen and underglaze. Gold lustre painted on. Fired to 705°C (1301°F).

This Shakespeare quotation is a long-standing favourite of Marcia's – she feels it is very relevant to clay and creativity. She pairs it here with other quotations from Shakespeare, by Walter Pater and Amy Lowell, also relating to the image of fire as an expression of creativity.

Dreams. Dimensions: 22 x 22 x 5 cm (8.7 x 8.7 x 2 in.). Thrown stoneware bowl. Raised letters Neuland-like. Incised letters in Neugebauer letter-form. Letters cut out and applied. Incised when reasonably stiff.

Large splash painted on with stoneware slip. Black underglaze sprayed when dry. Fired to 1260°C (2300°F).

Underglazes spattered on with toothbrush. Clear glaze brushed onto raised letters. Lichen glaze brushed onto some letters. Fired to 1025°C (1877°F).

Blue letters written with pen and underglaze. Splash painted with gold lustre. Fired to 705°C (1301°F).

From a series concerned with anti-war poetry. *Dreams* was inspired by a poem by Dana Burneet.

Bibliography

BIRKS, Tony, *The Complete Potter's Companion*, Conran Octopus, 1993.

BLANDINO, Betty, *Coiled Pottery*, A & C Black, 1984.

BROES, Jan, *Ein Stad vol Letters*, Brugge 2002.

CASTIN, Judy, *100 Great Calligraphy Tips*, Batsford, London.

COOPER, Emmanuel (Ed.), *Clays and Glazes*, Ceramic Review Publications, London.

FINK, Joanne & KASTIN, Judy, *Lettering Arts*, PBC Int. Inc. USA, 1993.

FLIGHT, Graham, *Keramik Manuel*, Collins, 1990.

GURTLER, Andre, *Experiments with Letter-form and Calligraphy*, 1997.

HALLIDAY, Peter, *Calligraphy – Art & Colour*, Batsford, London, 1994.

HAMER, Janet & Frank, *Potters Dictionary of Materials and Techniques*, A & C Black, 1993.

HARDY, Michael, *Handbuilding*, A & C Black, 2000.

LANE, Peter, *Studio Porcelain*, Pitman, 1980.

Modern Scribes and Lettering Artists, Studio Vista, 1980.

LANE, Peter, *Ceramic Form*, Collins, 1988.

LANE, Peter, *Contemporary Porcelain*, A & C Black/Chilton, 1995.

LANE, Peter, *Studio Ceramics*, Collins, 1983.

LASCHITZ, Elfriede, *Kalligraphie*, (*Calligraphy*), Novum Press, 1994.

MEDIAVILLA, Claude, *Calligraphy*, Scirpus Publications, 1996.

MORING, Annie, *Calligraphy Stroke by Stroke*, Simon & Schuster, Australia, 1995.

NOBLE Mary & MEHIGAN Janet, *The Calligraphers Companion*, Quarto, 1997

POTT, Gottfried, *The Music of Lettering*, 1995.

van der BRANT, Joke, *Kalligrafie in Europa*, Belgium, 1992.

RAU, Michael, KLOOS-RAU, Rosemarie, *Schreibschriften, Script Types*, Novum Press, Bruckmann, 1993.

VAN DIJK, Evert, *Kalligrafie*, Gaade Uitgevers.

WONDRAUSCH, Mary, *Mary Wondrausch on Slipware*, A & C Black, 2001.

Letter Arts Review, Published quarterly. Greensboro, USA.

John Neal Bookseller, P.O. Box 9986, Greensboro NC 27429 USA
info@johnnealbooks.com

(I can particularly recommend *Calligraphy*, *The Music of Lettering* and *Experiments with Letter-form and Calligraphy*.)

Suppliers List

Ceramic Suppliers, UK

Bath Potters' Supplies
Unit 18, Fourth Ave, Westfield Trading Estate, Radstock, Bath BA3 4XE
tel: 01761 411077
www.bathpotters.demon.co.uk

Brick House Ceramic Supplies
Cock Green, Felstead, Essex, CM6 3JE
tel: 01376 585655
www.members.lycos.co.uk/brickhouse

Ceramatech Ltd.
Units 16 & 17, Frontier Works, Queen Street, London N17 8JA
tel: 020 8885 4492

Potclays Ltd.
Brick Kiln Lane, Etruria, Stoke-on-Trent, Staffs. ST4 7BP
tel: 01782 219816
www.potclays.co.uk

Potterycrafts Ltd.
Campbell Road, Stoke-on-Trent, Staffs. ST4 4ET
tel: 01782 745000
www.potterycrafts.co.uk

Scarva Pottery Supplies
Unit 20, Scarva Rd Industrial Estate, Banbridge, Co. Down, BT32 3QD
tel: 018206 69699
www.scarvapottery.demon.co.uk

European Suppliers

Solargil
89520 Moutiers-en-Puisaye, France
tel: (00 33) 03 86 45 50 00
www.ceramique.com/solargil

Ceradel Socor
Z.I.N. 17 à 23 Rue Frédéric Bastiat
B.P. 1598, 87022 Limoges Cedex 9
France
tel: (00 33) 05 55 35 02 35
e mail: ceradel-socor@wanadoo.fr

Carl Jäger Tonindustriebedarf GmbH,
56202 Hilgert, Inden Erlen 4, Germany

SKG Keramische Farben GmbH
Hafenweg 26a, 48155 Münster, Germany

BSZ Keramikbedarf GmbH
Manderscheidtstr 90, 45141 Essen, Germany
tel: (00 49) 201 299 66

US Suppliers

American Art Clay Company
4717 W. 16th St, Indianapolis, IN 46222
tel: 317 244 6871
www.amaco.com

Axner Pottery Supplies
P.O. Box 621484, Oviedo, Florida 32762
tel: 800 843 7057
www.axner.com

Laguna Clay Co.
1440 Lomitas Avenue, City of Industry, CA 91746
tel: 800 843 7057
www.lagunaclay.com

Mile Hi Ceramics, Inc.
77 Lipan Street, Denver, Colorado 80223
tel: 303 825 4570
www.milehiceramics.com

Calligraphy Suppliers, UK

Blots
14 Lyndhurst Avenue, Preswich, Manchester M25 OGF
tel: 0161 720 6916

Cornelissen
105 Great Russell Street, London WC1B 3LA
tel: 020 7636 1045

Falkiner Fine Papers
76 Southampton Row, London WC1B 4AR
tel: 020 7831 1151

Scribblers
12 Witney Road, Pakefield, Lowestoft, Suffolk, NR33 7AW

North American Suppliers

Creative Calligraphy Equipment
239 Sheppard Avenue, Suite 202, Willowdale, Ontario M2N 3A8, Canada

Pendragon
862 Fairmount Avenue, St Paul Minesota 55105

Scribes Art Shop, Inc.
568 Jefferson Shopping Plaza, Port Jefferson Station, N.Y. 11776

European Suppliers

Comptoire des Ecritures
82, Rue Quincampoix, 75003 Paris, France

De Gouden Pluim
Vrijdagmarkt 12, 9000 Ghent, Belgium

Farbton
Blücherstr. 46, 6200 Wiesbaden, Germany

Index